T0213990

SpringerBriefs in Applied Sciences and Technology

Series Editor

Andreas Öchsner, Griffith School of Engineering, Griffith University, Southport, QLD, Australia

SpringerBriefs present concise summaries of cutting-edge research and practical applications across a wide spectrum of fields. Featuring compact volumes of 50 to 125 pages, the series covers a range of content from professional to academic.

Typical publications can be:

- A timely report of state-of-the art methods
- An introduction to or a manual for the application of mathematical or computer techniques
- A bridge between new research results, as published in journal articles
- A snapshot of a hot or emerging topic
- An in-depth case study
- A presentation of core concepts that students must understand in order to make independent contributions

SpringerBriefs are characterized by fast, global electronic dissemination, standard publishing contracts, standardized manuscript preparation and formatting guidelines, and expedited production schedules.

On the one hand, **SpringerBriefs in Applied Sciences and Technology** are devoted to the publication of fundamentals and applications within the different classical engineering disciplines as well as in interdisciplinary fields that recently emerged between these areas. On the other hand, as the boundary separating fundamental research and applied technology is more and more dissolving, this series is particularly open to trans-disciplinary topics between fundamental science and engineering.

Indexed by EI-Compendex, SCOPUS and Springerlink.

More information about this series at http://www.springer.com/series/8884

Tin-Chih Toly Chen

Advances in Fuzzy Group Decision Making

 Springer

Tin-Chih Toly Chen ⓘ
Department of Industrial Engineering
and Management
National Yang Ming Chiao Tung University
Hsinchu, Taiwan

ISSN 2191-530X   ISSN 2191-5318   (electronic)
SpringerBriefs in Applied Sciences and Technology
ISBN 978-3-030-86207-7   ISBN 978-3-030-86208-4   (eBook)
https://doi.org/10.1007/978-3-030-86208-4

This Springer imprint is published by the registered company Springer Nature Switzerland AG
The registered company address is: Gewerbestrasse 11, 6330 Cham, Switzerland

Contents

Chapter 1
Introduction to Fuzzy Group Decision-Making

1.1 Fuzzy Multiple Criteria Decision-Making

A multiple criteria decision-making (MCDM) problem may include both quantitative and qualitative criteria. However, sometimes it is difficult to evaluate the performance of an alternative when optimizing qualitative criteria [1]. To address this difficulty, the application of fuzzy logic to existing MCDM methods is a viable solution [2], which has given rise to a number of fuzzy MCDM methods, such as fuzzy decision trees [3], fuzzy inference systems (FISs) [4, 5], fuzzy Delphi method [6], fuzzy goal programming (FGP) [7], fuzzy analytic hierarchy process (FAHP) [8] and fuzzy analytic network process (FANP) [9], fuzzy multi-attribute utility theory (fuzzy MAUT) [10], fuzzy measuring attractiveness by a categorical based evaluation technique (fuzzy MACBETH) [11], fuzzy preference ranking organization method for enriched evaluation (fuzzy PROMETHEE) [12], fuzzy elimination and choice expressing reality method (fuzzy ELECTRE) [13], fuzzy technique for order of preference by similarity to ideal solution (fuzzy TOPSIS) [14], fuzzy VIšekriterijumskoKOmpromisnoRangiranje (fuzzy VIKOR) [15], and others. Figure 1.1 provides statistics on the popularity of these methods. There are also fuzzy MCDM method that come from fuzzy mathematics, such as fuzzy relationship (or preference) method [16], fuzzy synthetic evaluation [17], and weighted goals method [18]. These fuzzy MCDM methods can be divided into two major categories: fuzzy compromise MCDM methods and fuzzy outranking MCDM methods [19].

Taking fuzzy compromise MCDM methods as an example. A fuzzy compromise MCDM method comprises six steps [20] (Fig. 1.2). Fuzzy logic is usually applied to four of these steps: selecting factors and formulating criteria, selecting the aggregation method and determining the priorities of criteria, aggregation, and making a decision.

© The Author(s), under exclusive license to Springer Nature Switzerland AG 2021
T.-C. T. Chen, *Advances in Fuzzy Group Decision Making*,
SpringerBriefs in Applied Sciences and Technology,
https://doi.org/10.1007/978-3-030-86208-4_1

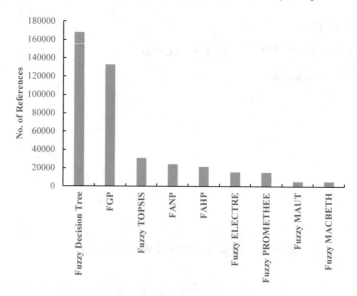

Fig. 1.1 The number of references about fuzzy decision-making methods from 2010 to 2021 (*Data source* Google Scholar)

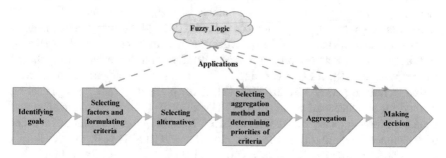

Fig. 1.2 The steps of a fuzzy compromise MCDM process

1.2   Fuzzy Group Decision-Making

To avoid personal bias and consider more viewpoints, a fuzzy MCDM process usually involves multiple decision-makers, which results in a fuzzy group decision-making (FGDM) problem [21]. In all steps, decision-makers discusses with each other. However, decision-makers usually have different opinions and make unequal judgments. Therefore, two additional steps are necessary:

(1)   Whether an overall consensus [22] exists among decision-makers should be checked.
(2)   The opinions, judgments, and/or decisions of decision-makers need to be aggregated [23].

If there is no overall consensus among decision-makers, the following treatments can be taken:

(1) Decision-makers with distinctive opinions, judgments, and/or decisions will be excluded [24].
(2) Decision-makers with distinctive opinions, judgments, and/or decisions will be asked to revise their opinions [24].
(3) Unequal authority levels will be assigned to decision-makers to weigh their opinions, judgments, and/or decisions [25].
(4) The partial consensus among most decision-makers is sought instead [26].

After decision-makers reach an overall (or partial) consensus, their opinions, judgments, and/or decisions can be aggregated. There are two ways of aggregation. An anterior (or first) aggregation method aggregates decision-makers' opinions (or judgments) before executing a step, while a posterior (or last) aggregation method allows decision-makers to execute a step individually and then aggregates their results [27], as illustrated in Fig. 1.3.

As a summary, existing FGDM methods can be classified into four categories according to the matrix in Fig. 1.4. For example, a Type I FGDM method aggregates decision-makers' opinions (or judgments) before applying a fuzzy compromise MCDM method such as FAHP.

Some FGDM examples are given below. Kahraman et al. [1] compared four FGDM methods for facility location selection: fuzzy relationship (or preference) method, fuzzy synthetic evaluation, weighted goals method, and FAHP [8]. Fuzzy methods are suitable for this problem because some criteria are qualitative, e.g., environmental regulations in a candidate location should be as loose as possible. The looseness of a regulation was difficult to accurately define or measure, so it could be suitably expressed as a fuzzy number. Three decision-makers made the decision collaboratively. Their opinions (or judgments) were aggregated in an anterior

Fig. 1.3 Anterior aggregation versus posterior aggregation

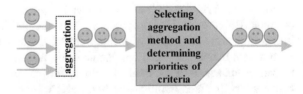

(a) An anterior (or first) aggregation method

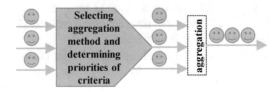

(b) A posterior (or last) aggregation method

Fig. 1.4 Categories of
FGDM methods

Category	Compromise Method	Outranking Method
Anterior Aggregation	I	II
Posterior Aggregation	III	IV

way. Common compromise (or aggregation) methods for this purpose include fuzzy weighted average (FWA) [28, 29], fuzzy geometric mean (FGM) [30], and fuzzy intersection (FI) [25]. Wang and Elhag [31] assessed the risks of a bridge using a FGDM method. The likelihood, consequence, and weight of a risk were subjectively evaluated or determined by each decision-maker. Then, FWA was applied to aggregate all decision-makers' evaluations. Cabrerizo et al. [32] pointed out that only after decision-makers reach a high degree of consensus can a joint decision be made. For this reason, they compared the advantages and disadvantages of various consensus measurement methods for a FGDM problem. Some methods evaluate the similarity (or proximity) between decision-makers' opinions (or preferences), others evaluate the similarity (or proximity) between their individual decisions. Turskis et al. [33] applied fuzzy Delphi method to choose the most influential person in the sustainable prevention of accidents in a construction project. In the fuzzy Delphi method, decision-makers were not allowed to communicate to maintain independence. Their opinions, in terms of fuzzy sets, were aggregated by averaging. Then, the aggregation result was provided to them to make possible adjustments. The FGDM process stopped when the aggregation result no longer changed significantly. The above examples highlights the variety of FGDM methods and the broad applicability of FGDM methods. An obvious research trend in this field is to use advanced types of fuzzy sets to further diversify the subjective judgments of decision-makers. For example, Boran et al. [34] proposed a FAHP method in which pairwise comparison results were expressed with intuitionistic fuzzy sets. An intuitionistic fuzzy set has both membership and non-membership functions to express the attitude of a decision-maker from different directions. Xu [35] proposed a similar method in which interval-valued intuitionistic fuzzy sets were adopted. In an interval-valued intuitionistic fuzzy set, the value of a membership function (or non-membership function) is an interval rather than a real number to allow for more vagueness and uncertainty. Other advanced types of fuzzy sets applied in FGDM methods include interval type-2 fuzzy sets [36], hesitant fuzzy sets [37], dual hesitant fuzzy sets [38], Pythagorean fuzzy sets [39], etc.

Existing FGDM methods are classified into the steps of a FGDM process in Table 1.1.

Table 1.1 Classification of existing FGDM methods into the steps of a fuzzy group decision-making process

Aggregating decision-makers' judgments	Deriving priorities of criteria	Aggregating fuzzy priorities	Checking the existence of consensus	Evaluating/Comparing alternatives	Aggregating decision-makers' evaluations
• Fuzzy arithmetic mean (FAM) • FWA • FGM • Fuzzy weighted geometric mean (FWGM) • Fuzzy best–worst method (fuzzy BWM) • Linguistic ordered weighted average (LOWA) • FI • Partial-consensus fuzzy intersection (PCFI) …	• FAM • FGM • Fuzzy extent analysis (FEA) • Fuzzy inverse of column sum (FICSM) • Piecewise linear FGM (PLFGM) • Alpha-cut operations (ACO) • Approximating ACO (xACO) • Fuzzy MACBETH …	• FAM • FWA • FGM • FWGM • FI • PCFI • Fuzzy weighted intersection (FWI) …	• FI • PCFI • FAM • Fuzzy similarity • Fuzzy proximity …	• FWA • Fuzzy MAUT • Fuzzy decision tree • FIS • Fuzzy TOPSIS • Fuzzy PROMETHEE • Fuzzy ELECTRE • Fuzzy VIKOR …	• Fuzzy relationships • FAM • FWA • FGM • FWGM • FI • PCFI …

1.3   Organization of This Book

This book is dedicated to introducing some progress of FGDM in theory and applications, including new concepts, execution procedures, solving methods, and various applications. In particular, this book emphasizes the introduction of fuzzy collaborative intelligence into FGDM to measure and enhance the consensus among decision-makers. The philosophy of fuzzy collaborative intelligence is to analyze a problem from diverse perspectives, so as to raise the chance that no relevant aspects of the problem will be ignored [40, 41]. From the view of fuzzy collaborative intelligence, in a fuzzy group decision-making system, some decision-makers (or decision-making units) with various backgrounds are trying to make a decision jointly through their collaboration [42–44]. Since they have different knowledge and points of view, they may apply various fuzzy decision-making methods to analyze the problem and make their choices. The key of such a system is that these decision-makers share and exchange their opinions, judgments, priorities, evaluations, and/or decisions with each other when making the final decision. In addition, existing FGDM methods usually average decision-makers' judgments, priorities, and/or evaluations to find their consensus, which is mathematically sound but may not be acceptable to all decision-makers [45]. In contrast, a fuzzy collaborative intelligence seeks the overlap of decision-makers' judgments, priorities, evaluations, and/or decisions to ensure the acceptability of the final decision [46].

In specific, the outline of the present book is structured as follows.

In the current chapter, first, the concept of fuzzy multiple criteria decision-making is introduced. Some prevalent fuzzy multiple criteria decision-making methods are also reviewed. Subsequently, the concept of FGDM is defined, which features the applications of fuzzy multiple criteria decision-making methods by multiple decision-makers in a collaborative manner. The steps of a FGDM process are detailed. In addition, some FGDM examples from the literate are also mentioned. Finally, existing FGDM methods are classified into the steps of a FGDM process.

Chapter 2, Fuzzy Group Decision-making Methods, provides an introduction of some FGDM methods. This chapter starts by dividing existing FGDM methods into several categories. Then, the FGDM methods of each category are introduced. Examples are also provided for some popular or newest FGDM methods. At the end of this chapter, we discuss how to apply FGDM methods amid the COVID-19 pandemic.

Chapter 3, Deriving the Priorities of Criteria, describes how to derive the fuzzy priorities of criteria, which is an essential task for a FGDM method. Most existing FGDM methods derive the (fuzzy) priorities of criteria from pairwise comparison results. However, owing to the complexity of fuzzy multiplication, it becomes a computationally intense task. To fulfill this task, some approximation methods are introduced in the first section of this chapter. The introduction of each method is accompanied by detailed numerical examples or programs. Subsequently, exact or near-exact methods for deriving the (fuzzy) priorities of criteria are reviewed. Most of the aforementioned methods are based on a ratio scale. Therefore, in the third section, a FGDM method based on an interval scale, fuzzy MACBETH, is detailed. Chapter 4, Consensus Measurement and Enhancement, deals with an important issue in FGDM, i.e., how to measure and enhance the consensus among decision-makers. This issue does not exist in a fuzzy decision-making problem that involves a single decision-maker. However, it is also a complex issue because the consensus among decision-makers can be measured at several time points. In this chapter, the traditional way to measure and enhance the consensus among decision-makers is first reviewed. Subsequently, from the view of fuzzy collaborative intelligence, how to measure and enhance the consensus among decision-makers is described.

In Chapter 5, Aggregation Mechanisms, some prevalent methods for aggregating decision-makers' judgments, priorities, evaluations, and/or decisions are first reviewed. Illustrative or numerical examples are also provided to deepen readers' understanding of these methods. Subsequently, the application of fuzzy collaborative intelligence methods to aggregate decision-makers' judgments, priorities, evaluations, and/or decisions is introduced.

References

1. C. Kahraman, D. Ruan, I. Doğan, Fuzzy group decision-making for facility location selection. Inf. Sci. **157**, 135–153 (2003)
2. J.M. Merigó, Fuzzy decision making with immediate probabilities. Comput. Ind. Eng. **58**(4), 651–657 (2010)
3. C. Olaru, L. Wehenkel, A complete fuzzy decision tree technique. Fuzzy Sets Syst. **138**(2), 221–254 (2003)
4. S.J. Fong, G. Li, N. Dey, R.G. Crespo, E. Herrera-Viedma, Composite Monte Carlo decision making under high uncertainty of novel coronavirus epidemic using hybridized deep learning and fuzzy rule induction. Appl. Soft Comput. **93**, 106282 (2020)
5. K. Govindan, H. Mina, B. Alavi, A decision support system for demand management in healthcare supply chains considering the epidemic outbreaks: a case study of coronavirus disease 2019 (COVID-19). Transp. Res. Part E: Logist. Transp. Rev. **138**, 101967 (2020)
6. A. Jafari, M. Jafarian, A. Zareei, F. Zaerpour, Using fuzzy Delphi method in maintenance strategy selection problem. J. Uncertain Syst. **2**(4), 289–298 (2008)
7. M.A. Parra, A.B. Terol, M.R. Uría, A fuzzy goal programming approach to portfolio selection. Eur. J. Oper. Res. **133**(2), 287–297 (2001)
8. L. Mikhailov, P. Tsvetinov, Evaluation of services using a fuzzy analytic hierarchy process. Appl. Soft Comput. **5**(1), 23–33 (2004)
9. L. Mikhailov, M.G. Singh, Fuzzy analytic network process and its application to the development of decision support systems. IEEE Trans. Syst. Man Cybernet., Part C (Applications and Reviews) **33**(1), 33–41 (2003)
10. A. Jimenez, A. Mateos, P. Sabio, Dominance intensity measure within fuzzy weight oriented MAUT: an application. Omega **41**(2), 397–405 (2013)
11. D. Dhouib, An extension of MACBETH method for a fuzzy environment to analyze alternatives in reverse logistics for automobile tire wastes. Omega **42**(1), 25–32 (2014)
12. Y.H. Chen, T.C. Wang, C.Y. Wu, Strategic decisions using the fuzzy PROMETHEE for IS outsourcing. Expert Syst. Appl. **38**(10), 13216–13222 (2011)
13. M. Sevkli, An application of the fuzzy ELECTRE method for supplier selection. Int. J. Prod. Res. **48**(12), 3393–3405 (2010)

14. C.C. Sun, A performance evaluation model by integrating fuzzy AHP and fuzzy TOPSIS methods. Expert Syst. Appl. **37**(12), 7745–7754 (2010)
15. H. Safari, Z. Faraji, S. Majidian, Identifying and evaluating enterprise architecture risks using FMEA and fuzzy VIKOR. J. Intell. Manuf. **27**(2), 475–486 (2016)
16. J.M. Blin, Fuzzy relations in group decision theory. J. Cybernet. **4**, 17–22 (1974)
17. D.Y. Chang, Applications of the extent analysis method on fuzzy AHP. Eur. J. Oper. Res. **95**, 649–655 (1996)
18. R.R. Yager, Fuzzy decision-making including unequal objectives. Fuzzy Sets Syst. **1**, 87–95 (1978)
19. A. Ishizaka, P. Nemery, *Multi-Criteria Decision Analysis Methods and Software* (Wiley, 2013)
20. D. Rajapakse, MCDM: multiple criteria decision making—a boring introduction (2017). https://medium.com/@dileesha/mcdm-multiple-criteria-decision-making-a-boring-int roduction-1e0062f2e48
21. B. Vahdani, S.M. Mousavi, H. Hashemi, M. Mousakhani, R. Tavakkoli-Moghaddam, A new compromise solution method for fuzzy group decision-making problems with an application to the contractor selection. Eng. Appl. Artif. Intell. **26**(2), 779–788 (2013)
22. J. Liu, F.T. Chan, Y. Li, Y. Zhang, Y. Deng, A new optimal consensus method with minimum cost in fuzzy group decision. Knowl.-Based Syst. **35**, 357–360 (2012)
23. R. Yuan, J. Tang, F. Meng, Linguistic intuitionistic fuzzy group decision making based on aggregation operators. Int. J. Fuzzy Syst. **21**(2), 407–420 (2019)
24. H. Gao, Y. Ju, E.D.S. Gonzalez, W. Zhang, Green supplier selection in electronics manufacturing: an approach based on consensus decision making. J. Cleaner Product. 118781 (2019)
25. T. Chen, Y.C. Lin, A fuzzy-neural system incorporating unequally important expert opinions for semiconductor yield forecasting. Int. J. Uncertain. Fuzz. Knowl.-Based Syst. **16**(01), 35–58 (2008)
26. T. Chen, A hybrid fuzzy and neural approach with virtual experts and partial consensus for DRAM price forecasting. Int. J. Innov. Comput. Inf. Control **8**(1), 583–597 (2012)
27. E. Roghanian, J. Rahimi, A. Ansari, Comparison of first aggregation and last aggregation in fuzzy group TOPSIS. Appl. Math. Model. **34**(12), 3754–3766 (2010)
28. T. Chen, Evaluating the sustainability of a smart technology application to mobile health care—the FGM-ACO-FWA approach. Complex Intell. Syst. **6**, 109–121 (2020)
29. Y.C. Wang, T. Chen, Y.L. Yeh, Advanced 3D printing technologies for the aircraft industry: a fuzzy systematic approach for assessing the critical factors. Int. J. Adv. Manuf. Technol. **105**, 4059–4069 (2019)
30. J.J. Buckley, Fuzzy hierarchical analysis. Fuzzy Sets Syst. **17**, 233–247 (1985)
31. Y.M. Wang, T.M. Elhag, A fuzzy group decision making approach for bridge risk assessment. Comput. Ind. Eng. **53**(1), 137–148 (2007)
32. F.J. Cabrerizo, J.M. Moreno, I.J. Pérez, E. Herrera-Viedma, Analyzing consensus approaches in fuzzy group decision making: advantages and drawbacks. Soft. Comput. **14**(5), 451–463 (2010)
33. Z. Turskis, S. Dzitac, A. Stankiuviene, R. Šukys, A fuzzy group decision-making model for determining the most influential persons in the sustainable prevention of accidents in the construction SMEs. Int. J. Comput. Commun. Control **14**(1), 90–106 (2019)
34. F.E. Boran, S. Genç, M. Kurt, D. Akay, A multi-criteria intuitionistic fuzzy group decision making for supplier selection with TOPSIS method. Expert Syst. Appl. **36**(8), 11363–11368 (2009)
35. Z. Xu, A method based on distance measure for interval-valued intuitionistic fuzzy group decision making. Inf. Sci. **180**(1), 181–190 (2010)
36. J. Qin, X. Liu, W. Pedrycz, A multiple attribute interval type-2 fuzzy group decision making and its application to supplier selection with extended LINMAP method. Soft. Comput. **21**(12), 3207–3226 (2017)
37. L.W. Lee, S.M. Chen, Fuzzy decision making and fuzzy group decision making based on likelihood-based comparison relations of hesitant fuzzy linguistic term sets 1. J. Intell. Fuzzy Syst. **29**(3), 1119–1137 (2015)

38. D. Yu, D.F. Li, J.M. Merigo, Dual hesitant fuzzy group decision making method and its application to supplier selection. Int. J. Mach. Learn. Cybern. **7**(5), 819–831 (2016)
39. S. Zeng, X. Peng, T. Baležentis, D. Streimikiene, Prioritization of low-carbon suppliers based on Pythagorean fuzzy group decision making with self-confidence level. Economic Research-Ekonomska Istraživanja **32**(1), 1073–1087 (2019)
40. W. Pedrycz, Collaborative architectures of fuzzy modeling. Lect. Notes Comput. Sci. **5050**, 117–139 (2008)
41. T.C.T. Chen, K. Honda, *Fuzzy Collaborative Forecasting and Clustering: Methodology, System Architecture, and Applications* (Springer, Switzerland AG, 2019)
42. W. Pedrycz, P. Rai, A multifaceted perspective at data analysis: a study in collaborative intelligent agents. IEEE Trans. Syst. Man Cybernet., Part B (Cybernetics) **38**(4), 1062–1072 (2008)
43. W. Pedrycz, Collaborative fuzzy clustering. Pattern Recognit. Lett. **23**, 1675–1686 (2002)
44. S. Mitra, H. Banka, W. Pedrycz, Rough–fuzzy collaborative clustering. IEEE Trans. Syst. Man Cybernet., Part B (Cybernetics) **36**(4), 795–805 (2006)
45. F. Herrera, E. Herrera-Viedma, A model of consensus in group decision making under linguistic assessments. Fuzzy Sets Syst. **78**(1), 73–87 (1996)
46. T. Chen, Y.C. Wang, M.C. Chiu, Assessing the robustness of a factory amid the COVID-19 pandemic: a fuzzy collaborative intelligence approach. Healthcare **8**, 481 (2020)

Chapter 2
Fuzzy Group Decision-Making Methods

Bozdağ et al. [1] classified fuzzy group decision-making methods into seven categories:

- Fuzzy simple additive weighting methods;
- Fuzzy analytic hierarchy process (FAHP) methods;
- Fuzzy conjunctive/disjunctive methods;
- Fuzzy outranking methods;
- Maximin methods;
- Fuzzy technique for order preference by similarity to ideal solution (fuzzy TOPSIS) methods;
- Linguistic methods.

However, to be applicable for a fuzzy group decision-making process, these methods still lack the following features:

- A method for aggregating the judgments, decisions, and/or preferences of decision makers;
- A mechanism for measuring and enhancing the consensus among decision makers.

Prevalent fuzzy group decision-making methods include fuzzy weighted averaging (FWA) [2, 3], intuitionistic fuzzy weighted averaging (IFWA) [4], linguistic ordered weighted average (LOWA) [5], fuzzy TOPSIS [6], intuitionistic fuzzy TOPSIS [4], FAHP [7], fuzzy best–worst method (fuzzy BWM) [8], fuzzy extent analysis (FEA) [9], fuzzy geometric mean (FGM) [10], alpha-cut operations (ACO) [11], fuzzy intersection (FI) [12], partial-consensus FI (PCFI) [13], and others. These methods are applied to different stages of a fuzzy group decision-making process.

T.-C. T. Chen, *Advances in Fuzzy Group Decision Making*,
SpringerBriefs in Applied Sciences and Technology,
https://doi.org/10.1007/978-3-030-86208-4_2

2.1 Fuzzy Outranking Methods

Blin's fuzzy relations. Kahraman et al. [6] compared four fuzzy group decision-making methods for choosing a suitable facility location. The first method is Blin's fuzzy relations [14], in which each decision maker first ranks all alternatives individually. Then, all decision makers' ranking results are aggregated by calculating the percentage such that one alternative is preferred to the other, based on which a fuzzy relation is developed. After specifying a threshold for the membership to the fuzzy relation, only qualified preferences are left, based on which the ranking of all alternatives is generated.

Similar methods were also applied in Bozdağ et al. [1] for choosing suitable computer integrated manufacturing systems.

Büyüközkan et al. [15] designed rules to generate a customer's preferences for various quality functions. Then, the preferences of multiple customers were aggregated using the LOWA operator. In this way, the most preferred quality functions should be realized by a product design.

In Capuano et al. [16], decision makers first express their preferences for alternatives. Then, an iterative procedure was followed so that decision makers adjusted their preferences by considering those of others that were considered more authoritative. Such a procedure obviously facilitated the convergence of decision makers' preferences for alternatives.

2.2 FWA

FWA. Turskis et al. [2] specified the priorities of criteria by decision makers and then averaged them. Subsequently, FWA was applied to evaluate the overall performance of each alternative.

Wang and Elhag [17] assessed the risks of bridges, in which decision makers evaluated the likelihood, consequence, and priority of each risk with linguistic terms that were mapped to fuzzy numbers. The evaluation results by decision makers were averaged. Then, each risk was assessed by multiplying its likelihood to consequence. Finally, the overall risk was derived using FWA.

IFWA-intuitionistic fuzzy TOPSIS. Boran et al. [4] dealt with a facility location selection problem, in which the performances of alternatives in optimizing each criterion were evaluated by decision makers in linguistic terms that were mapped to intuitionistic fuzzy numbers. Decision makers had unequal authority levels (or weights). Therefore, their evaluation results were aggregated using IFWA [18]. The priorities of criteria were determined in the same way. Finally, intuitionistic fuzzy TOPSIS was applied to evaluate and compare the overall performances of alternatives.

2.3 Fuzzy TOPSIS

Banaeian et al. [19] aimed to choose a suitable green supplier, in which decision makers subjectively assigned priorities to criteria. The priorities assigned by decision makers were averaged, and then fed into three uncertain evaluation methods including fuzzy TOPSIS, fuzzy VIšekriterijumskoKOmpromisnoRangiranje (fuzzy VIKOR), and gray relational analysis (GRA) for comparison. The priorities of criteria are multiplied to the normalized performance in fuzzy TOPSIS, but to the distance of an alternative to the best (or the worst) solution.

Fuzzy TOPSIS is usually applied to evaluate the overall performance of an alternative. First, the performance of an alternative in optimizing each criterion is normalized using fuzzy distributive normalization [20] as

$$
\begin{aligned}
\tilde{\rho}_{qi} &= \frac{\tilde{p}_{qi}}{\sqrt{\sum_{r=1}^{Q} \tilde{p}_{ri}^2}} \\
&= \frac{1}{\sqrt{1(+)\sum_{r \neq q}\left(\frac{\tilde{p}_{ri}}{\tilde{p}_{qi}}\right)^2}} \\
&= \left(\frac{1}{\sqrt{1+\sum_{r \neq q}\left(\frac{p_{ri3}}{p_{qi1}}\right)^2}}, \frac{1}{\sqrt{1+\sum_{r \neq q}\left(\frac{p_{ri2}}{p_{qi2}}\right)^2}}, \frac{1}{\sqrt{1+\sum_{r \neq q}\left(\frac{p_{ri1}}{p_{qi3}}\right)^2}}\right)
\end{aligned} \tag{2.1}
$$

where $\tilde{p}_{qi}$ is the performance of the qth alternative in optimizing the ith criterion; $\tilde{\rho}_{qi}$ is the normalized performance. (+) denotes fuzzy addition. Subsequently, the fuzzy priorities of criteria are multiplied to the normalized performances:

$$
\begin{aligned}
\tilde{s}_{qi} &= \tilde{w}_i(\times)\tilde{\rho}_{qi} \\
&= (w_{i1}\rho_{qi1},\ w_{i2}\rho_{qi2},\ w_{i3}\rho_{qi3})
\end{aligned} \tag{2.2}
$$

$(\times)$ denotes fuzzy multiplication. The fuzzy ideal (zenith) point and the fuzzy anti-ideal (nadir) point are specified, respectively, as [20]

$$
\begin{aligned}
\tilde{\Lambda}^+ &= \{\tilde{\Lambda}_i^+\} \\
&= \{\max_q \tilde{s}_{qi}\} \\
&= \{\max_q s_{qi1},\ \max_q s_{qi2},\ \max_q s_{qi3}\}
\end{aligned} \tag{2.3}
$$

$$
\begin{aligned}
\tilde{\Lambda}^- &= \{\tilde{\Lambda}_i^-\} \\
&= \{\min_q \tilde{s}_{qi}\} \\
&= \{\min_q s_{qi1},\ \min_q s_{qi2},\ \min_q s_{qi3}\}
\end{aligned} \tag{2.4}
$$

The fuzzy distance from each alternative to the two reference points are measured, respectively, as [20]

$$
\tilde{d}_q^+ = \sqrt{\sum_{i=1}^{n} \max(\tilde{\Lambda}_i^+ (-)\tilde{s}_{qi},\ 0)^2}
$$

$$
= \left(\sqrt{\sum_{i=1}^{n} \max(\Lambda_{i1}^+ - s_{qi3},\ 0)^2},\ \sqrt{\sum_{i=1}^{n} \max(\Lambda_{i2}^+ - s_{qi2},\ 0)^2},\ \sqrt{\sum_{i=1}^{n} \max(\Lambda_{i3}^+ - s_{qi1},\ 0)^2} \right) \quad (2.5)
$$

$$
\tilde{d}_q^- = \sqrt{\sum_{i=1}^{n} \max(\tilde{s}_{qi}(-)\tilde{\Lambda}_i^-,\ 0)^2}
$$

$$
= \left(\sqrt{\sum_{i=1}^{n} \max(s_{qi1} - \Lambda_{i3}^-,\ 0)^2},\ \sqrt{\sum_{i=1}^{n} \max(s_{qi2} - \Lambda_{i2}^-,\ 0)^2},\ \sqrt{\sum_{i=1}^{n} \max(s_{qi3} - \Lambda_{i1}^-,\ 0)^2} \right) \quad (2.6)
$$

$(-)$ denotes fuzzy subtraction. The difference between fuzzy TOPSIS and FWA is that the formula for measuring a distance adopted in FTOPSIS is quadratic (i.e., the Euclidean distance), while that adopted in FWA is linear [21].

Finally, the fuzzy closeness of the alternative is obtained as [20]

$$
\tilde{C}_q = \frac{\tilde{d}_q^-}{\tilde{d}_q^+ (+)\tilde{d}_q^-}
$$

$$
= \frac{1}{\frac{\tilde{d}_q^+}{\tilde{d}_q^-}(+)1}
$$

$$
= \left(\frac{1}{\frac{d_{q3}^+}{d_{q1}^-}+1},\ \frac{1}{\frac{d_{q2}^+}{d_{q2}^-}+1},\ \frac{1}{\frac{d_{q1}^+}{d_{q3}^-}+1} \right) \quad (2.7)
$$

The fuzzy closeness can be defuzzified using the COG formula [22] as

$$
D(\tilde{C}_q) = \frac{\int_0^1 x \mu_{\tilde{C}_q}(x)dx}{\int_0^1 \mu_{\tilde{C}_q}(x)dx}
$$

$$
= \frac{C_{q1} + C_{q2} + C_{q3}}{3} \quad (2.8)
$$

The overall performance of an alternative is higher if its defuzzified closeness is higher.

Example 2.1 Fuzzy TOPSIS is applied to evaluate and compare the overall performances of six alternatives. The performances of these alternatives in optimizing five criteria are summarized in Table 2.1. First, the performances are normalized. The results are shown in Table 2.2.

Assume the fuzzy priorities of criteria are

$$
\tilde{w}_1 = (0.13,\ 0.27,\ 0.46)
$$

Table 2.1 Performances of six alternatives

q	$\tilde{p}_{q1}$	$\tilde{p}_{q2}$	$\tilde{p}_{q3}$	$\tilde{p}_{q4}$	$\tilde{p}_{q5}$
1	(1.5, 2.5, 3.5)	(4, 5, 5)	(4, 5, 5)	(4, 5, 5)	(4, 5, 5)
2	(0, 0, 1)	(0, 0, 1)	(4, 5, 5)	(0, 0, 1)	(0, 0, 1)
3	(4, 5, 5)	(3, 4, 5)	(4, 5, 5)	(3, 4, 5)	(4, 5, 5)
4	(0, 1, 2)	(3, 4, 5)	(0, 0, 1)	(3, 4, 5)	(0, 0, 1)
5	(1.5, 2.5, 3.5)	(1.5, 2.5, 3.5)	(4, 5, 5)	(3, 4, 5)	(0, 0, 1)
6	(3, 4, 5)	(3, 4, 5)	(0, 0, 1)	(4, 5, 5)	(4, 5, 5)

Table 2.2 Normalized performances

q	$\tilde{\rho}_{q1}$	$\tilde{\rho}_{q2}$	$\tilde{\rho}_{q3}$	$\tilde{\rho}_{q4}$	$\tilde{\rho}_{q5}$
1	(0.18, 0.34, 0.56)	(0.39, 0.56, 0.68)	(0.41, 0.5, 0.59)	(0.37, 0.51, 0.61)	(0.48, 0.58, 0.66)
2	(0, 0, 0.18)	(0, 0, 0.15)	(0.41, 0.5, 0.59)	(0, 0, 0.13)	(0, 0, 0.14)
3	(0.48, 0.68, 0.81)	(0.3, 0.45, 0.64)	(0.41, 0.5, 0.59)	(0.29, 0.4, 0.58)	(0.48, 0.58, 0.66)
4	(0, 0.14, 0.35)	(0.3, 0.45, 0.64)	(0, 0, 0.12)	(0.29, 0.4, 0.58)	(0, 0, 0.14)
5	(0.18, 0.34, 0.63)	(0.15, 0.28, 0.51)	(0.41, 0.5, 0.59)	(0.29, 0.4, 0.65)	(0, 0, 0.17)
6	(0.38, 0.54, 0.74)	(0.3, 0.45, 0.64)	(0, 0, 0.12)	(0.37, 0.51, 0.61)	(0.48, 0.58, 0.66)

$$\tilde{w}_2 = (0.25, \ 0.47, \ 0.65)$$
$$\tilde{w}_3 = (0.02, \ 0.04, \ 0.12)$$
$$\tilde{w}_4 = (0.08, \ 0.16, \ 0.36)$$
$$\tilde{w}_5 = (0.03, \ 0.06, \ 0.12)$$

After multiplying the fuzzy priorities to the normalized performances, the weighted normalized performances are obtained (see Table 2.3). From these results, the fuzzy ideal (zenith) point and the fuzzy anti-ideal (nadir) point are specified (see Table 2.4).

Subsequently, the distances from each alternative to the two reference points are measured, based on which the fuzzy closeness of the alternative is calculated.

Table 2.3 Weighted normalized performances

q	$\tilde{s}_{q1}$	$\tilde{s}_{q2}$	$\tilde{s}_{q3}$	$\tilde{s}_{q4}$	$\tilde{s}_{q5}$
1	(0.02, 0.09, 0.25)	(0.1, 0.27, 0.44)	(0.01, 0.02, 0.07)	(0.03, 0.08, 0.22)	(0.01, 0.03, 0.08)
2	(0, 0, 0.08)	(0, 0, 0.1)	(0.01, 0.02, 0.07)	(0, 0, 0.05)	(0, 0, 0.02)
3	(0.06, 0.18, 0.37)	(0.08, 0.21, 0.42)	(0.01, 0.02, 0.07)	(0.02, 0.06, 0.21)	(0.01, 0.03, 0.08)
4	(0, 0.04, 0.16)	(0.08, 0.21, 0.42)	(0, 0, 0.01)	(0.02, 0.06, 0.21)	(0, 0, 0.02)
5	(0.02, 0.09, 0.29)	(0.04, 0.13, 0.34)	(0.01, 0.02, 0.07)	(0.02, 0.06, 0.23)	(0, 0, 0.02)
6	(0.05, 0.14, 0.34)	(0.08, 0.21, 0.42)	(0, 0, 0.01)	(0.03, 0.08, 0.22)	(0.01, 0.03, 0.08)

Table 2.4 Fuzzy ideal (zenith) point and fuzzy anti-ideal (nadir) point

Reference point	$\tilde{\Lambda}_1^*$	$\tilde{\Lambda}_2^*$	$\tilde{\Lambda}_3^*$	$\tilde{\Lambda}_4^*$	$\tilde{\Lambda}_5^*$
Fuzzy ideal (zenith) point	(0.06, 0.18, 0.37)	(0.1, 0.27, 0.44)	(0.01, 0.02, 0.07)	(0.03, 0.08, 0.23)	(0.01, 0.03, 0.08)
Fuzzy anti-ideal (nadir) point	(0, 0, 0.08)	(0, 0, 0.1)	(0, 0, 0.01)	(0, 0, 0.05)	(0, 0, 0.02)

Table 2.5 Distances from each alternative to the two reference points

q	$\tilde{d}_q^+$	$\tilde{d}_q^+$	$\tilde{C}_q$	$D(\tilde{C}_q)$
1	(0, 0.09, 0.54)	(0, 0.29, 0.57)	(0.01, 0.77, 1)	0.59
2	(0, 0.33, 0.63)	(0, 0.02, 0.15)	(0, 0.06, 0.98)	0.35
3	(0, 0.06, 0.53)	(0, 0.29, 0.6)	(0, 0.84, 1)	0.61
4	(0, 0.16, 0.57)	(0, 0.22, 0.49)	(0, 0.58, 1)	0.53
5	(0, 0.16, 0.58)	(0, 0.17, 0.51)	(0, 0.51, 1)	0.50
6	(0, 0.07, 0.54)	(0, 0.27, 0.59)	(0, 0.8, 1)	0.60

The results are summarized in Table 2.5. The defuzzification result of the fuzzy closeness using COG is also presented in this table. Obviously, Alternative #3 is the best performing alternative.

Torfi et al. [23] also compared the effectiveness of FAHP with that of fuzzy TOPSIS for a fuzzy multiple criteria decision-making problem. In FAHP, fuzzy arithmetic mean (FAM), rather than the prevalent FGM or FEA, was applied to derive the fuzzy priorities of criteria. In addition, in fuzzy TOPSIS,

(1) The fuzzy ideal (zenith) point and the fuzzy anti-ideal (nadir) point were set to 1 and 0, respectively.
(2) The distances from an alternative to the two reference points are crisp:

$$d_q^+ = \sqrt{\frac{1}{3} \sum_{i=1}^n ((\Lambda_{i1}^+ - s_{qi1})^2 + (\Lambda_{i2}^+ - s_{qi2})^2 + (\Lambda_{i3}^+ - s_{qi3})^2)} \qquad (2.9)$$

$$d_q^- = \sqrt{\frac{1}{3} \sum_{i=1}^n ((\Lambda_{i1}^- - s_{qi1})^2 + (\Lambda_{i2}^- - s_{qi2})^2 + (\Lambda_{i3}^- - s_{qi3})^2)} \qquad (2.10)$$

As a result, the closeness of an alternative is also crisp and does not need to be defuzzified. However, to a fuzzy group decision-making process, crisp-valued distances and closeness are not conducive to the aggregation of decision makers' evaluation results. Decision makers can only aggregate their judgments or the fuzzy priorities of criteria derived by them.

2.4 FAHP and Fuzzy BWM

The combination of fuzzy logic with analytic hierarchy process (AHP) is called FAHP that provides fuzzy-valued weights or priorities that are more flexible than crisp values [24]. Unlike AHP, in FAHP decision makers' judgments are modeled by fuzzy values, which provides a lot of space for the subsequent aggregation in a fuzzy group decision-making process. A number of the previous studies applied FAHP methods [7]. Although these methods are called FAHP methods, FAHP is used to derive the fuzzy priorities of criteria. Then, another method needs to be applied to evaluate and compare the overall performances of alternatives.

Gnanavelbabu and Arunagiri [25] proposed a FAHP method to rank several sources of MUDA, in which FEA was applied to derive the priorities of criteria in crisp values. In this way, decision makers should aggregate their judgments before applying the FAHP method.

Wang et al. [26] proposed a FAHP method to select the most suitable maintenance strategy. To simplify the required calculation, the priorities of criteria were derived as crisp values, which is the same as in the FEA method. In this way, the membership of the relative priority of two criteria in the corresponding pairwise comparison result was greater than a threshold, as illustrated in Fig. 2.1.

However, the traditional consistency measure became invalid. To solve this problem, a new consistency measure was proposed:

$$\gamma = e^{-\max_{i,j}\left(\mu_{\tilde{a}_{ij}}\left(\frac{w_i}{w_j}\right)\right)}$$ (2.11)

As a result, for a fuzzy group decision-making problem, it is better to aggregate decision makers' judgments before applying the proposed FAHP method.

Junior et al. [27] investigated a supplier selection problem, to which both FAHP and fuzzy TOPSIS were applied for comparison. When FAHP was applied, decision makers' judgments were aggregated using FAM, rather than the common FGM, which might be problematic. Subsequently, FEA was applied to derive the fuzzy priorities of criteria. In contrast, decision makers reached a consensus before applying fuzzy TOPSIS to make a joint decision.

Fig. 2.1 Requirement for crisp priorities of criteria

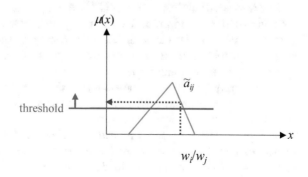

FEA-FWA. In Kahraman et al. [6], multiple decision makers compared the relative priorities of criteria for choosing a suitable facility location jointly using FEA. Subsequently, FWA was applied to evaluate the overall performance of an alternative. The priorities of criteria derived using FEA were crisp. There was no need for aggregation.

Fuzzy synthetic evaluation. Kahraman et al. [6] applied a fuzzy synthetic evaluation method for a facility location selection problem, in which decision makers evaluated the performance of an alternative in optimizing a criterion jointly with multiple linguistic terms and memberships, e.g., "Excellent" with a membership of 0.7 and "Superior" with a membership of 0.4. Therefore, there was no need to aggregate decision makers' judgments. Subsequently, the priorities of criteria were determined using an analytic hierarchy process (AHP) method [28], and applied to aggregate the performances of an alternative in optimizing all criteria. The alternative with the highest membership at a higher level was selected.

Yager's weighted goals method. In Yager's weighted goals method [6], decision makers evaluated the performance of an alternative in optimizing a criterion jointly with multiple linguistic terms and memberships. Then, the priorities of criteria are determined using AHP, which become the powers of memberships. The minimal membership of an alternative in optimizing all criteria determines its overall performance, based on which alternatives are ranked. Decision makers have reached a consensus before applying this method, so there is no need for aggregation.

FGM-FEA. Wang et al. [29] chose critical factors for choosing suitable three-dimensional (3D) printing technologies for the aircraft industry. First, decision makers' judgments were aggregated using FGM. Then, FEA was applied to derive the priorities of critical factors.

Fuzzy BWM. Pishdar et al. [8] proposed a fuzzy best–worst method (fuzzy BWM) to choose a suitable hub airport. In the fuzzy BWM method, each decision maker first chose the most important and the least important criteria. Then, the other criteria were compared with the two criteria in pairs. In this way, the number of pairwise comparisons required could be reduced. Subsequently, a mathematical programming problem, like in Measuring Attractiveness by a Categorical-Based Evaluation TecHnique (MACBETH) [30], was solved to derive the priorities of criteria. Criterion priorities derived by all decision makers were averaged. Finally, the overall performance of an alternative was evaluated using FWA.

Fuzzy logarithmic least squares (FLLS). Zhang and Chu [31] proposed an FLLS method to derive the priorities from a fuzzy judgment matrix. In addition, the fuzzy judgment matrixes of decision makers were averaged. Then, the deviation from the fuzzy judgment matrix of each decision maker to the average was minimized to enhance the consensus among decision makers.

Fuzzy analytic hierarchy network (FANP). Promentilla et al. [32] proposed a FANP method to evaluate contaminated site remedial countermeasures. The FANP method took the dependencies between criteria and alternatives into consideration. In addition, it derived the fuzzy priorities of criteria and evaluated the overall performances of alternatives at the same time. In the FANP method, a pairwise comparison result was replaced with the weighted sum of its α cuts:

$$\tilde{A} = \begin{vmatrix} 1 & \tilde{a}_{12} & 1/\tilde{a}_{31} \\ 1/\tilde{a}_{12} & 1 & 1/\tilde{a}_{32} \\ \tilde{a}_{31} & \tilde{a}_{32} & 1 \end{vmatrix}$$

$$\rightarrow A(\alpha) = \begin{vmatrix} 1 & \omega a_{12}^L + (1-\omega)a_{12}^R & 1/(\omega a_{31}^L + (1-\omega)a_{31}^R) \\ 1/(\omega a_{12}^L + (1-\omega)a_{12}^R) & 1 & 1/(\omega a_{32}^L + (1-\omega)a_{32}^R) \\ \omega a_{31}^L + (1-\omega)a_{31}^R & \omega a_{32}^L + (1-\omega)a_{32}^R & 1 \end{vmatrix}$$

$$(2.12)$$

$\alpha = 0, 0.1, \ldots, 1$. As a result, the problem became a crisp analytic hierarchy network (ANP) problem for each α level, which considerably simplified the required calculation. Although the problem is a fuzzy group decision-making problem, how to aggregate decision makers' judgments or decisions was not disclosed, which meant that decision makers reached a consensus before applying the proposed FANP method.

Guaranteed-consensus FAHP. A challenge facing FAHP methods for a fuzzy group decision-making problem is that the fuzzy priorities of criteria derived by different decision makers may not overlap, implying a lack of consensus among them. To overcome this challenge, Chen [33] proposed a guaranteed-consensus FAHP method. In the guaranteed-consensus FAHP method, first, the ranges of triangular fuzzy numbers (TFNs) for expressing pairwise comparison results are widened, so that the derived fuzzy priorities of criteria will overlap, as illustrated in Fig. 2.2.

After deriving the fuzzy priorities of criteria, a systematic procedure is followed, so that the ranges of these TFNs can be tightened to increase the precision of fuzzy priorities while still maintaining the consensus, as illustrated in Fig. 2.3.

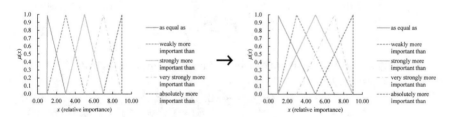

Fig. 2.2 Widening the TFNs for pairwise comparison results

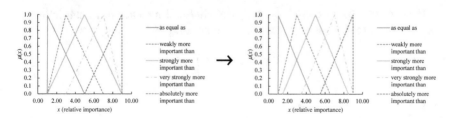

Fig. 2.3 Tightening the TFNs for pairwise comparison results

Multi-belief FAHP. Lin and Chen [34] believed that a decision maker has multiple beliefs about the relative priority of a criterion, but is not sure which one to choose, and is forced to aggregate these beliefs into a single judgment. Therefore, it is essential to restore these beliefs in order to make the right decision. To this end, in their methodology, a nonlinear programming problem is solved to decompose a decision maker's judgment matrix into several single-belief judgment matrixes that are more consistent than the original judgment matrix and represent diversified points of view regarding the relative priorities of criteria.

Subsequently, Chen and Lin [35] extended their methodology to a fuzzy decision-making case in which diversified 3D printers were to be chosen. A fuzzy mixed integer nonlinear programming (FMINLP) problem was solved to decompose a fuzzy judgment matrix into several fuzzy sub-judgment matrixes that were diverse and more consistent than the original fuzzy judgment matrix. Based on each fuzzy sub-judgment matrix, a suitable 3D printer could be chosen. In this way, diverse 3D printers could be chosen for a manufacturer.

Assume the fuzzy judgment matrix of a decision maker is indicated by $\tilde{\mathbf{A}}$. The decision maker may have multiple beliefs about the relative priorities of criteria, which are mapped to multiple fuzzy sub-judgment matrixes $[\tilde{\mathbf{A}}(m); m = 1 \sim M]$. These fuzzy sub-judgment matrixes can be consolidated to form the original fuzzy judgment matrix [36]:

$$\tilde{\mathbf{A}} := \frac{\sum_{m=1}^{M} \tilde{\mathbf{A}}(m)}{M} \tag{2.13}$$

Namely,

$$\tilde{a}_{ij} := \frac{\sum_{m=1}^{M} \tilde{a}_{ij}(m)}{M} \forall a_{ij} > 1 \tag{2.14}$$

All fuzzy sub-judgment matrixes meet the basic requirements [25]:

$$det(\tilde{\mathbf{A}}(m)(-)\tilde{\lambda}(m)\mathbf{I}) = 0 \tag{2.15}$$

$$(\tilde{\mathbf{A}}(m)(-)\tilde{\lambda}(m)\mathbf{I})(\times)\tilde{\mathbf{x}}(m) = 0 \tag{2.16}$$

It is expected that by considering a single belief at a time, each fuzzy sub-judgment matrix will be more consistent than the original fuzzy judgment matrix:

$$\widetilde{CR}(\tilde{\mathbf{A}}(k)) \leq \widetilde{CR}(\tilde{\mathbf{A}}) \tag{2.17}$$

where $\widetilde{CR}$ is fuzzy consistency ratio [25]:

$$\widetilde{CR} = \frac{\tilde{\lambda}_{\max} - n}{n - 1}/RI \tag{2.18}$$

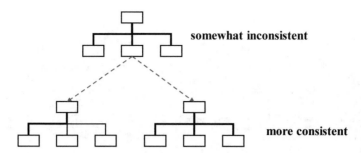

Fig. 2.4 Decomposing the original fuzzy judgment matrix into several fuzzy sub-judgment matrixes

where RI is the random consistency index [25]. This is illustrated in Fig. 2.4.

Example 2.2 A decision maker constructs the following fuzzy judgment matrix:

$$\tilde{\mathbf{A}} = \begin{vmatrix} 1 & (3,\ 5,\ 7) & - & - & - \\ - & 1 & - & - & - \\ (1,\ 3,\ 5) & (5,\ 7,\ 9) & 1 & (2,\ 4,\ 6) & (2,\ 4,\ 6) \\ (1,\ 3,\ 5) & (3,\ 5,\ 7) & - & 1 & - \\ (1,\ 3,\ 5) & (2,\ 4,\ 6) & - & (3,\ 5,\ 7) & 1 \end{vmatrix}$$

$\widetilde{CR}(\tilde{\mathbf{A}}) = (0,\ 0.17,\ 6.89)$. $\tilde{\mathbf{A}}$ can be decomposed into the following two fuzzy sub-judgment matrixes:

$$\tilde{\mathbf{A}}(1) = \begin{vmatrix} 1 & (1,\ 2,\ 4) & - & - & - \\ - & 1 & - & - & - \\ (2,\ 4,\ 6) & (3,\ 5,\ 9) & 1 & (1,\ 1,\ 3) & (1,\ 1,\ 3) \\ (3,\ 5,\ 7) & (6,\ 8,\ 9) & - & 1 & - \\ (2,\ 4,\ 6) & (3,\ 5,\ 7) & - & (5,\ 7,\ 9) & 1 \end{vmatrix}$$

$$\tilde{\mathbf{A}}(2) = \begin{vmatrix} 1 & (6,\ 8,\ 9) & - & - & - \\ - & 1 & - & - & - \\ (1,\ 2,\ 4) & (7,\ 9,\ 9) & 1 & (5,\ 7,\ 9) & (5,\ 7,\ 9) \\ (1,\ 1,\ 3) & (1,\ 2,\ 4) & - & 1 & - \\ (1,\ 2,\ 4) & (1,\ 3,\ 5) & - & (1,\ 3,\ 5) & 1 \end{vmatrix}$$

such that $\tilde{\mathbf{A}} := (\tilde{\mathbf{A}}(1)(+)\tilde{\mathbf{A}}(2))/2$. The two fuzzy sub-judgment matrixes are more consistent:

$$\widetilde{CR}(\tilde{\mathbf{A}}(1)) = (0,\ 0.15,\ 4.57)$$
$$\widetilde{CR}(\tilde{\mathbf{A}}(2)) = (0,\ 0.12,\ 5.61)$$

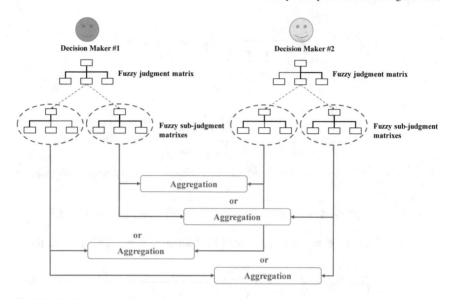

Fig. 2.5 Flexible aggregation of the fuzzy sub-judgment matrixes of decision makers

The multi-belief FAHP method can be applied to a fuzzy group decision-making case. Since fuzzy sub-judgment matrixes are more consistent than the original fuzzy judgment matrixes, aggregating the fuzzy sub-judgment matrixes, rather than the fuzzy judgment matrixes, of decision makers can help them make a more flexible and reliable decision, as illustrated in Fig. 2.5.

Approximating ACO. For a FAHP problem, ACO can derive the exact membership functions of the fuzzy maximal eigenvalue and the fuzzy priorities of criteria. However, it is time-consuming. To tackle this difficulty, Chen et al. [37] proposed an approximating ACO method that fits the membership function of the fuzzy maximal eigenvalue (or a fuzzy priority) with two nonlinear functions on the left-hand and right-hand sides, respectively, during the ACO process. In this way, it is no longer necessary to enumerate all possible combinations of the α cuts of pairwise comparison results, thereby significantly enhancing the efficiency of ACO. For a fuzzy group decision-making process, aggregating the fuzzy priorities of criteria derived by decision makers based on these exact results is expected to generate a more trustable decision.

2.5 Fuzzy Collaborative Intelligence Methods

In fuzzy collaborative intelligence methods, the consensus among decision makers is sought using FI at various stages of a fuzzy group decision-making process.

FGM-PCFI. Wang and Chen [38] proposed a partial-consensus posterior aggregation FAHP method for a supplier selection problem. Each decision maker first derived the fuzzy priorities of criteria using FGM individually. Then, the PCFI operator was applied to aggregate the fuzzy priorities derived by all decision makers. In contrast, most other studies averaged the fuzzy priorities derived by decision makers. In addition, "partial consensus" referred to only aggregating the fuzzy priorities derived by most decision makers, which helped to deal with the lack of overall consensus.

ACO-FI-FWA. Chen et al. [11] proposed a fuzzy collaborative intelligence approach to evaluate the suitability of smart health practice. First, decision makers applied ACO to derive the exact priorities of criteria. Then, FI was applied to aggregate the priorities derived by all decision makers. If the FI result was an empty set, decision makers needed to modify their judgments. Otherwise, based on the aggregation result, FWA was applied to evaluate the suitability of smart health practice.

ACO-FI-fuzzy TOPSIS. Lin et al. [39] proposed another fuzzy collaborative intelligence method, in which ACO and FI were applied to derive the priorities of criteria for each decision maker and aggregate the priorities derived by all decision makers, respectively. Finally, the aggregated priorities were fed into the fuzzy TOPSIS method [26] to evaluate the suitability of a smart technology application for fall detection.

Piecewise linear FGM. To enhance the efficiency of ACO, Wu et al. [40] proposed a piecewise linear FGM approach to approximate the membership function of a fuzzy priority with two piecewise linear functions.

2.6 Fuzzy Group Decision-Making Methods Amid the COVID-19 Pandemic

There are many uncertainties and risks in making decisions amid the COVID-19 pandemic [41, 42]. Therefore, it is even more necessary to concentrate the wisdom of multiple decision makers. To this end, some fuzzy or probabilistic decision-making methods [43–45] have been proposed. For example, Wu et al. [46] proposed a fuzzy collaborative intelligence-based fuzzy analytic hierarchy process (FAHP) approach to evaluate and compare 15 intervention strategies in response to the COVID-19 pandemic. In their methodology, each decision maker applies the FGM method to derive the fuzzy priorities of criteria. Then, the layered partial-consensus approach [47] is applied to aggregate the derived fuzzy priorities by most decision makers. Finally, the generalized fuzzy weighted assessment approach is proposed to evaluate the effectiveness of an intervention strategy for tackling the COVID-19 pandemic. Chen and Lin [48] proposed the FAHP method for comparing various smart and automation technology applications to ensure the long-term operation of a factory amid the COVID-19 pandemic. Chen et al. [49] proposed a fuzzy collaborative intelligence method to evaluate the robustness of a factory to the COVID-19 pandemic. Fuzzy intersection and partial-consensus fuzzy intersection were applied to aggregate decision makers' evaluation results, depending on the number of experts who have reached a consensus. Fong et al. [41] conducted Monte Carlo simulations to extrapolate several time series data of the COVID-19 pandemic. Then, they built a deep learning network to predict these time series. Based on these predictions, a fuzzy inference system (FIS) was developed to analyze the trend of the COVID-19 pandemic. Wu et al. [50] applied machine learning technologies [51, 52] to assess the severity risk of an incoming COVID-19 patient, thereby making decisions in the allocation of medical resources. Govindan et al. [53] developed an FIS for assisting demand management in a healthcare supply chain. To investigate the impact of the COVID-19 pandemic on family investment decisions, Yue et al. [54] applied linear probability and probit models. The experimental results showed that households who knew someone infected with COVID-19 lost confidence in the economy and might become risk-averse. Burlea-Schiopoiu and Ferhati [55] applied the structural equation modeling (SEM) method to identify factors critical to the performance of a healthcare sector, thereby defining key performance indexes. The COVID-19 pandemic has impacted many industries. Yu et al. [56] proposed a similar SEM method to identify factors that may influence people's fear of missing out, thereby guiding people's decisions to repost news related to the COVID-19 pandemic on social media. Lystad et al. [57] fitted seasonal autoregressive integrated moving average (ARIMA) models to predict the utilization of manual therapy services in Australia. The decline is expected upon utilization, providing valuable information for service providers and governments to consider in making responsive decisions. The supply chain disruption caused by the coronavirus disease 2019 (COVID-19) pandemic has forced many manufacturers to look for alternative suppliers. To fulfill this demand, Chen et al. [58] propose a calibrated FGM (cFGM)-fuzzy FTOPSIS-fuzzy weighted intersection (FWI) approach to choose a suitable alternative supplier in the COVID-19 pandemic. In their methodology, first, cFGM is proposed to accurately derive the fuzzy priorities of criteria. Subsequently, each decision maker applies fuzzy TOPSIS to compare the overall performances of alternative suppliers in the COVID-19 pandemic. Finally, the FWI operator is used to aggregate the comparison results by all decision makers, for which a decision maker's authority level is set to a value proportional to the consistency of his/her pairwise comparison results.

References

1. C.E. Bozdağ, C. Kahraman, D. Ruan, Fuzzy group decision making for selection among computer integrated manufacturing systems. Comput. Ind. **51**(1), 13–29 (2003)
2. Z. Turskis, S. Dzitac, A. Stankiuviene, R. Šukys, A fuzzy group decision-making model for determining the most influential persons in the sustainable prevention of accidents in the construction SMEs. Int. J. Comput. Commun. Control **14**(1), 90–106 (2019)

3. F. Liu, J.M. Mendel, Aggregation using the fuzzy weighted average as computed by the Karnik-Mendel algorithms. IEEE Trans. Fuzzy Syst. **16**(1), 1–12 (2008)
4. F.E. Boran, S. Genç, M. Kurt, D. Akay, A multi-criteria intuitionistic fuzzy group decision making for supplier selection with TOPSIS method. Expert Syst. Appl. **36**(8), 11363–11368 (2009)
5. F. Herrera, E. Herrera-Viedma, A model of consensus in group decision making under linguistic assessments. Fuzzy Sets Syst. **78**(1), 73–87 (1996)
6. C. Kahraman, D. Ruan, I. Doğan, Fuzzy group decision-making for facility location selection. Inf. Sci. **157**, 135–153 (2003)
7. G. Zheng, N. Zhu, Z. Tian, Y. Chen, B. Sun, Application of a trapezoidal fuzzy AHP method for work safety evaluation and early warning rating of hot and humid environments. Saf. Sci. **50**(2), 228–239 (2012)
8. M. Pishdar, F. Ghasemzadeh, J. Antuchevičienė, A mixed interval type-2 fuzzy best-worst MACBETH approach to choose hub airport in developing countries: case of Iranian passenger airports. Transport **34**(6), 639–651 (2019)
9. D.Y. Chang, Applications of the extent analysis method on fuzzy AHP. Eur. J. Oper. Res. **95**, 649–655 (1996)
10. J.J. Buckley, Fuzzy hierarchical analysis. Fuzzy Sets Syst **17**(3), 233–247 (1985)
11. T.C.T. Chen, Y.C. Wang, Y.C. Lin, H.C. Wu, H.F. Lin, A fuzzy collaborative approach for evaluating the suitability of a smart health practice. Mathematics **7**(12), 1180 (2019)
12. T. Chen, Y.C. Lin, A fuzzy-neural system incorporating unequally important expert opinions for semiconductor yield forecasting. Int. J. Uncertain. Fuzz. Knowl.-Based Syst. **16**(01), 35–58 (2008)
13. T. Chen, A hybrid fuzzy and neural approach with virtual experts and partial consensus for DRAM price forecasting. Int. J. Innov. Comput. Inf. Control. **8**, 583–597 (2012)
14. J.M. Blin, Fuzzy relations in group decision theory. J. Cybernet. **4**, 17–22 (1974)
15. G. Büyüközkan, O. Feyzioğlu, D. Ruan, Fuzzy group decision-making approach to multiple preference formats in quality function deployment. Comput. Ind. **58**(5), 392–402 (2007)
16. N. Capuano, F. Chiclana, H. Fujita, E. Herrera-Viedma, V. Loia, Fuzzy group decision making with incomplete information guided by social influence. IEEE Trans. Fuzzy Syst. **26**(3), 1704–1718 (2017)
17. Y.M. Wang, T.M. Elhag, A fuzzy group decision making approach for bridge risk assessment. Comput. Ind. Eng. **53**(1), 137–148 (2007)
18. Z.S. Xu, Intuitionistic fuzzy aggregation operators. IEE Trans. Fuzzy Syst. **15**(6), 1179–1187 (2007)
19. N. Banaeian, H. Mobli, B. Fahimnia, I.E. Nielsen, M. Omid, Green supplier selection using fuzzy group decision making methods: a case study from the agri-food industry. Comput. Oper. Res. **89**, 337–347 (2018)
20. Y. Nazar, R.A.P. Lovian, D.C. Raharjo, C.N. Rosyidi, Supplier selection and order allocation using TOPSIS and linear programming method at Pt. Sekarlima Surakarta. AIP Conf. Proc. **2097**(1), 030050 (2019)
21. Y.C. Lin, T. Chen, L.C. Wang, Integer nonlinear programming and optimized weighted-average approach for mobile hotel recommendation by considering travelers' unknown preferences. Oper. Res. Int. J. **18**(3), 625–643 (2018)
22. E. Van Broekhoven, B. De Baets, Fast and accurate center of gravity defuzzification of fuzzy system outputs defined on trapezoidal fuzzy partitions. Fuzzy Sets Syst. **157**(7), 904–918 (2006)
23. F. Torfi, R.Z. Farahani, S. Rezapour, Fuzzy AHP to determine the relative weights of evaluation criteria and Fuzzy TOPSIS to rank the alternatives. Appl. Soft Comput. **10**(2), 520–528 (2010)
24. P.J.M. Van Laarhoven, W. Pedrycz, A fuzzy extension of Saaty's priority theory. Fuzzy Sets Syst. **11**(1–3), 229–241 (1983)
25. A. Gnanavelbabu, P. Arunagiri, Ranking of MUDA using AHP and Fuzzy AHP algorithm. Mater. Today Proc. **5**(5), 13406–13412 (2018)
26. L. Wang, J. Chu, J. Wu, Selection of optimum maintenance strategies based on a fuzzy analytic hierarchy process. Int. J. Prod. Econ. **107**(1), 151–163 (2007)

27. F.R.L. Junior, L. Osiro, L.C.R. Carpinetti, A comparison between Fuzzy AHP and Fuzzy TOPSIS methods to supplier selection. Appl. Soft Comput. **21**, 194–209 (2014)
28. T.L. Saaty, Decision making with the analytic hierarchy process. Int. J. Serv. Sci. **1**(1), 83–98 (2008)
29. Y.C. Wang, T. Chen, Y.L. Yeh, Advanced 3D printing technologies for the aircraft industry: a fuzzy systematic approach for assessing the critical factors. Int. J. Adv. Manuf. Technol. **105**, 4059–4069 (2019)
30. D. Dhouib, Fuzzy Macbeth method to analyze alternatives in automobile tire wastes reverse logistics, in *2013 International Conference on Advanced Logistics and Transport* (2013), pp. 321–326
31. Z. Zhang, X. Chu, Fuzzy group decision-making for multi-format and multi-granularity linguistic judgments in quality function deployment. Expert Syst. Appl. **36**(5), 9150–9158 (2009)
32. M.A.B. Promentilla, T. Furuichi, K. Ishii, N. Tanikawa, A fuzzy analytic network process for multi-criteria evaluation of contaminated site remedial countermeasures. J. Environ. Manage. **88**(3), 479–495 (2008)
33. T.C.T. Chen, Guaranteed-consensus posterior-aggregation fuzzy analytic hierarchy process method. Neural Comput. Appl. **32**, 1–12 (2020)
34. Y.C., Lin, T. Chen, A multibelief analytic hierarchy process and nonlinear programming approach for diversifying product designs: smart backpack design as an example. Proc. Inst. Mech. Eng., Part B: J. Eng. Manufact. **234**(6–7), 1044–1056 (2020)
35. T.C.T. Chen, Y.C. Lin, Diverse three-dimensional printing capacity planning for manufacturers. Robot. Comput.-Integrat. Manufact. **67**, 102052 (2021)
36. M. Hanss, *Applied Fuzzy Arithmetic* (Springer, Berlin Heidelberg, 2005)
37. T. Chen, Y.C. Lin, M.C. Chiu, Approximating alpha-cut operations approach for effective and efficient fuzzy analytic hierarchy process analysis. Appl. Soft Comput. **85**, 105855 (2019)
38. Y.C. Wang, T.C.T. Chen, A partial-consensus posterior-aggregation FAHP method—supplier selection problem as an example. Mathematics **7**(2), 179 (2019)
39. Y.C. Lin, Y.C. Wang, T.C.T. Chen, H.F. Lin, Evaluating the suitability of a smart technology application for fall detection using a fuzzy collaborative intelligence approach. Mathematics **7**(11), 1097 (2019)
40. H.-C. Wu, T. Chen, C.-H. Huang, A piecewise linear FGM approach for efficient and accurate FAHP analysis: Smart backpack design as an example. Mathematics **8**, 1319 (2020)
41. S.J. Fong, G. Li, N. Dey, R.G. Crespo, E. Herrera-Viedma, Composite Monte Carlo decision making under high uncertainty of novel coronavirus epidemic using hybridized deep learning and fuzzy rule induction. Appl. Soft Comput. **93**, 106282 (2020)
42. P. Coulthard, Dentistry and coronavirus (COVID-19)-moral decision-making. Br. Dent. J. **228**, 503–505 (2020)
43. P. Melin, J.C. Monica, D. Sanchez, O. Castillo, Multiple ensemble neural network models with fuzzy response aggregation for predicting COVID-19 time series: the case of Mexico. Healthcare **8**, 181 (2020)
44. M. Toğaçar, B. Ergen, Z. Cömert, COVID-19 detection using deep learning models to exploit Social Mimic Optimization and structured chest X-ray images using fuzzy color and stacking approaches. Comput. Biol. Med. **93**, 103805 (2020)
45. Y.L. Fu, K.C. Liang, Fuzzy logic programming and adaptable design of medical products for the COVID-19 anti-epidemic normalization. Comput. Methods Programs Biomed. **197**, 105762 (2020)
46. H.C. Wu, Y.C. Wang, T.C.T. Chen, Assessing and comparing COVID-19 intervention strategies using a varying partial consensus fuzzy collaborative intelligence approach. Mathematics **8**, 1725 (2020)
47. T.C.T. Chen, H.C. Wu, Forecasting the unit cost of a DRAM product using a layered partial-consensus fuzzy collaborative forecasting approach. Complex Int. Syst. **6**, 479–492 (2020)
48. T. Chen, C.-W. Lin, Smart and automation technologies for ensuring the long-term operation of a factory amid the COVID-19 pandemic: an evolving fuzzy assessment approach. Int. J. Adv. Manuf. Technol. **111**, 3545–3558 (2020)

49. T. Chen, Y.C. Wang, M.C. Chiu, Assessing the robustness of a factory amid the COVID-19 pandemic: a fuzzy collaborative intelligence approach. Healthcare **8**, 481 (2020)
50. G. Wu, P. Yang, Y. Xie, H.C. Woodruff, X. Rao, J. Guiot, A.-N. Frix, R. Louis, M. Moutschen, J. Li et al., Development of a clinical decision support system for severity risk prediction and triage of COVID-19 patients at hospital admission: an international multicentre study. Eur. Respir. J. **56**, 2001104 (2020)
51. M.C. Chiu, T.C.T. Chen, Assessing sustainable effectiveness of the adjustment mechanism of a ubiquitous clinic recommendation system. Health Care Manag. Sci. **23**, 239–248 (2020)
52. T.C.T. Chen, M.C. Chiu, Mining the preferences of patients for ubiquitous clinic recommendation. Health Care Manag. Sci. **23**, 173–184 (2020)
53. K. Govindan, H. Mina, B. Alavi, A decision support system for demand management in healthcare supply chains considering the epidemic outbreaks: a case study of coronavirus disease 2019 (COVID-19). Transp. Res. Part E: Logist. Transp. Rev. **138**, 101967 (2020)
54. P. Yue, A.G. Korkmaz, H. Zhou, Household financial decision making amidst the COVID-19 pandemic. Emerg. Mark. Financ. Trade **56**, 2363–2377 (2020)
55. A. Burlea-Schiopoiu, K. Ferhati, The managerial implications of the key performance indicators in healthcare sector: a cluster analysis. Healthcare **9**, 19 (2020)
56. S.C. Yu, H.R. Chen, A.C. Liu, H.Y. Lee, Toward COVID-19 information: infodemic or fear of missing out? Healthcare **8**, 550 (2020)
57. R.P. Lystad, B.T. Brown, M.S. Swain, R.M. Engel, Impact of the COVID-19 pandemic on manual therapy service utilization within the Australian private healthcare setting. Healthcare **8**, 558 (2020)
58. T. Chen, Y.C. Wang, H.C. Wu, Analyzing the impact of vaccine availability on alternative supplier selection amid the COVID-19 pandemic: a cFGM-FTOPSIS-FWI approach. Healthcare **9**(1), 71 (2021)

Chapter 3
Deriving the Priorities of Criteria

3.1 Approximation Methods

A common practice in fuzzy multiple-criteria decision making is to compare the priorities of criteria in pairs [1–5]. Specifically, the relative priority of a criterion over another is expressed in linguistic terms such as "as equal as", "weakly more important than", "strongly more important than", "very strongly more important than", "absolutely more important than", etc. These linguistic terms are usually mapped to triangular fuzzy numbers (TFNs) within [1, 9] [6], as illustrated in Table 3.1.

Based on pairwise comparison results, a fuzzy judgment matrix $\tilde{\mathbf{A}}_{n \times n} = [\tilde{a}_{ij}]$ is constructed, in which

$$
\tilde{a}_{ji} = \begin{cases} 1 & if \quad j = i \\ 1/\tilde{a}_{ij} & otherwise \end{cases} \tag{3.1}
$$

The fuzzy eigenvalue and eigenvector of $\tilde{\mathbf{A}}$, indicated with $\tilde{\lambda}$ and $\tilde{\mathbf{x}}$ respectively, are derived by solving the following equations:

$$
det(\tilde{\mathbf{A}}(-)\tilde{\lambda}\mathbf{I}) = 0 \tag{3.2}
$$

$$
(\tilde{\mathbf{A}}(-)\tilde{\lambda}\mathbf{I})(\times)\tilde{\mathbf{x}} = 0 \tag{3.3}
$$

where $(-)$ and $(\times)$ denote fuzzy subtraction and multiplication, respectively. Subsequently, the relative priority (or weight) of criterion i can be derived as

$$
\tilde{w}_i = \frac{\tilde{x}_i}{\sum_{k=1}^{n} \tilde{x}_k} \tag{3.4}
$$

© The Author(s), under exclusive license to Springer Nature Switzerland AG 2021
T.-C. T. Chen, *Advances in Fuzzy Group Decision Making*,
SpringerBriefs in Applied Sciences and Technology,
https://doi.org/10.1007/978-3-030-86208-4_3

Table 3.1 Linguistic terms for pairwise comparison

Linguistic term	TFN
As equal as	(1, 1, 3)
As equal as or weakly more important than	(1, 2, 4)
Weakly more important than	(1, 3, 5)
Weakly or strongly more important than	(2, 4, 6)
Strongly more important than	(3, 5, 7)
Strongly or very strongly more important than	(4, 6, 8)
Very strongly more important than	(5, 7, 9)
Very strongly or absolutely more important than	(6, 8, 9)
Absolutely more important than	(7, 9, 9)

However, Eqs. (3.2) and (3.3) involve many fuzzy multiplication operations. As a result, $\tilde{\lambda}$, $\tilde{x}$ and $\tilde{w}_i$ are no longer TFNs [7]. It is also a computationally intensive task to derive their exact values [8]. For solving these problems, various approximation techniques have been proposed.

3.1.1 Fuzzy Geometric Mean (FGM)

Fuzzy geometric mean (FGM) is one of the most widely applied methods for approximating the fuzzy priorities of criteria in FAHP problems. The FGM method approximates the values of fuzzy priorities as [9]

$$\tilde{w}_i \cong \frac{\sqrt[n]{\prod_{j=1}^{n} \tilde{a}_{ij}}}{\sum_{k=1}^{n} \sqrt[n]{\prod_{j=1}^{n} \tilde{a}_{kj}}} \tag{3.5}$$

To this end, the following theorem is helpful.

Theorem 3.1 [10]

$$w_{i1} \cong \frac{1}{1 + \sum_{k \neq i} \frac{\sqrt[n]{\prod_{j=1}^{n} a_{kj3}}}{\sqrt[n]{\prod_{j=1}^{n} a_{ij1}}}} \tag{3.6}$$

$$w_{i2} \cong \frac{1}{1 + \sum_{k \neq i} \frac{\sqrt[n]{\prod_{j=1}^{n} a_{kj2}}}{\sqrt[n]{\prod_{j=1}^{n} a_{ij2}}}} \tag{3.7}$$

$$w_{i3} \cong \frac{1}{1 + \sum_{k \neq i} \frac{\sqrt[n]{\prod_{j=1}^{n} a_{kj1}}}{\sqrt[n]{\prod_{j=1}^{n} a_{ij3}}}} \tag{3.8}$$

Based on (3.8), the fuzzy maximal eigenvalue can be estimated as

$$\tilde{\lambda}_{\max} = \frac{\tilde{A}(\times)\tilde{w}}{\tilde{w}}$$

$$= \frac{1}{n} \sum_{i=1}^{n} \left(\frac{\sum_{j=1}^{n} (\tilde{a}_{ij}(\times)\tilde{w}_j)}{\tilde{w}_i} \right) \tag{3.9}$$

Theorem 3.2 [10]

$$\lambda_{\max,1} \cong 1 + \frac{1}{n} \sum_{i=1}^{n} \sum_{j\neq i} \frac{a_{ij1} w_{j1}}{w_{i3}} \tag{3.10}$$

$$\lambda_{\max,2} \cong 1 + \frac{1}{n} \sum_{i=1}^{n} \sum_{j\neq i} \frac{a_{ij2} w_{j2}}{w_{i2}} \tag{3.11}$$

$$\lambda_{\max,3} \cong 1 + \frac{1}{n} \sum_{i=1}^{n} \sum_{j\neq i} \frac{a_{ij3} w_{j3}}{w_{i1}} \tag{3.12}$$

The consistency among pairwise comparison results can be evaluated in terms of fuzzy consistency ratio ($\widetilde{CR}$) as

$$\widetilde{CR} = \frac{\tilde{\lambda}_{\max} - n}{n - 1} / RI$$

$$= \left(\max\left(\frac{\lambda_{\max,1} - n}{(n-1)RI}, \ 0 \right), \ \frac{\lambda_{\max,2} - n}{(n-1)RI}, \ \frac{\lambda_{\max,3} - n}{(n-1)RI} \right) \tag{3.13}$$

where RI is random consistency index [11] (see Table 3.2). $\widetilde{CR}$ should be less than 0.1 for a small problem. When the size of a judgment matrix is large, the requirement for $\widetilde{CR}$ can be relaxed to being less than 0.3.

Ignoring the dependency between the dividend and divisor of Eq. (3.5) results in a simplified version indicated with FGMi. However, the approximated value of w_{i3} may be greater than 1.

Example 3.1 A decision maker constructs the following fuzzy judgment matrix:

Table 3.2 Random consistency index

n	1	2	3	4	5	6	7	8	9	10
RI	0	0	0.58	0.9	1.12	1.24	1.32	1.41	1.45	1.49

$$\tilde{\mathbf{A}} = \begin{vmatrix} 1 & (2,\ 4,\ 6) & (3,\ 5,\ 7) \\ 1/(2,\ 4,\ 6) & 1 & (1,\ 2,\ 4) \\ 1/(3,\ 5,\ 7) & 1/(1,\ 2,\ 4) & 1 \end{vmatrix}$$

According to Theorem 3.1, the value of $\tilde{w}_1$ can be approximated as

$$w_{11} \cong \frac{1}{1 + \frac{\sqrt[3]{1/2 \cdot 1 \cdot 4}}{\sqrt[3]{1 \cdot 2 \cdot 3}} + \frac{\sqrt[3]{1/3 \cdot 1/1 \cdot 1}}{\sqrt[3]{1 \cdot 2 \cdot 3}}} = 0.48$$

$$w_{12} \cong \frac{1}{1 + \frac{\sqrt[3]{1/4 \cdot 1 \cdot 2}}{\sqrt[3]{1 \cdot 4 \cdot 5}} + \frac{\sqrt[3]{1/5 \cdot 1/2 \cdot 1}}{\sqrt[3]{1 \cdot 4 \cdot 5}}} = 0.68$$

$$w_{13} \cong \frac{1}{1 + \frac{\sqrt[3]{1/6 \cdot 1 \cdot 1}}{\sqrt[3]{1 \cdot 6 \cdot 7}} + \frac{\sqrt[3]{1/7 \cdot 1/4 \cdot 1}}{\sqrt[3]{1 \cdot 6 \cdot 7}}} = 0.80$$

Similarly, the values of $\tilde{w}_2$ and $\tilde{w}_3$ are derived as

$$\tilde{w}_2 = (0.12,\ 0.20,\ 0.37)$$

$$\tilde{w}_3 = (0.07,\ 0.12,\ 0.23)$$

Subsequently, Theorem 3.2 is applied to approximate the value of $\tilde{\lambda}_{max}$ as

$$\lambda_{max,1} \cong 1 + \frac{1}{3}\left(\frac{2 \cdot 0.12}{0.8} + \frac{3 \cdot 0.07}{0.8} + \frac{1/6 \cdot 0.48}{0.37} + \frac{1 \cdot 0.07}{0.37} + \frac{1/7 \cdot 0.48}{0.23} + \frac{1/4 \cdot 0.12}{0.23} \right)$$
$$= 1.45$$

$$\lambda_{max,2} \cong 3.02$$

$$\lambda_{max,3} \cong 10.62$$

As a result the consistency ratio is evaluated as

$$\widetilde{CR} \cong (0,\ 0.02,\ 6.57)$$

3.1.2   Calibrated FGM (cFGM) Method

To improve the accuracy of deriving the priorities of criteria, the calibrated FGM approach (cFGM) was proposed by Chen and Wang [12] as follows:

(1) Approximate the value of $\tilde{w}_i$ using the traditional FGM method as
(2) Treat $\tilde{\mathbf{A}}$ as a crisp one by considering only the core of $\tilde{a}_{ij}$. In this situation, the relative priority of criterion i is derived exactly using an eigen analysis as $w_i^{(c)}$.
(3) Calibrate the value of $\tilde{w}_i$ in the following manner:

$$\tilde{w}_i \rightarrow \tilde{w}_i + w_i^{(c)} - w_{i2}$$
$$= (w_{i1} + w_i^{(c)} - w_{i2}, \ w_i^{(c)}, \ w_{i3} + w_i^{(c)} - w_{i2}) \tag{3.14}$$

Example 3.2 The fuzzy judgment matrix of a decision maker is given below:

$$\tilde{\mathbf{A}} = \begin{vmatrix} 1 & (3, 5, 7) & - & - & - \\ - & 1 & - & - & - \\ (3, 5, 7) & (7, 9, 9) & 1 & (2, 4, 6) & (1, 3, 5) \\ (1, 3, 5) & (3, 5, 7) & - & 1 & - \\ (1, 3, 5) & (2, 4, 6) & - & (1, 3, 5) & 1 \end{vmatrix}$$

The value of $\tilde{w}_3$ derived using the traditional FGM method is $(0.258, 0.495, 0.681)$. After treating $\tilde{\mathbf{A}}$ as a crisp matrix, the value of $w_3^{(c)}$ is derived using an eigen analysis as 0.482. Therefore,

$$\tilde{w}_3 \rightarrow \tilde{w}_3 + w_3^{(c)} - w_{32}$$
$$= (0.245, \ 0.482, \ 0.668)$$

as shown in Fig. 3.1.

Fig. 3.1 Comparison of the fuzzy priorities derived using FGM and cFGM

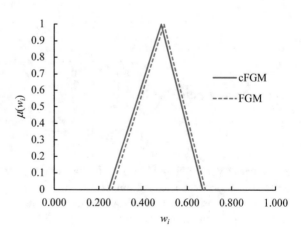

3.1.3 Fuzzy Extent Analysis (FEA)

The fuzzy extent analysis (FEA) method starts from evaluating the fuzzy synthetic extent of each criterion as [13]

$$\tilde{s}_i = \frac{\sum_{j=1}^n \tilde{a}_{ij}}{\sum_{h=1}^n \sum_{j=1}^n \tilde{a}_{hj}} \tag{3.15}$$

Theorem 3.3 [14, 15]

$$s_{i1} = \frac{1}{1 + \sum_{h \neq i} \frac{\sum_{j=1}^n a_{hj3}}{\sum_{j=1}^n a_{ij1}}} \tag{3.16}$$

$$s_{i2} = \frac{1}{1 + \sum_{h \neq i} \frac{\sum_{j=1}^n a_{hj2}}{\sum_{j=1}^n a_{ij2}}} \tag{3.17}$$

$$s_{i3} = \frac{1}{1 + \sum_{h \neq i} \frac{\sum_{j=1}^n a_{hj1}}{\sum_{j=1}^n a_{ij3}}} \tag{3.18}$$

Then, the priority of the criterion is set to its minimal degree of being the maximum:

$$w_i = \min_{j \neq i}(V(\tilde{s}_i \geq \tilde{s}_j)) \tag{3.19}$$

where

$$V(\tilde{s}_i \geq \tilde{s}_j) = \begin{cases} 1 & \text{if} \quad s_{i2} \geq s_{j2} \\ 0 & \text{if} \quad s_{i3} \leq s_{j1} \\ \frac{s_{j1} - s_{i3}}{(s_{i2} - s_{i3}) - (s_{j2} - s_{j1})} & \text{otherwise} \end{cases} \tag{3.20}$$

$\{w_i\}$ needs to be normalized:

$$w_i \rightarrow \frac{w_i}{\sum_{j=1}^n w_j} \tag{3.21}$$

In this way, the obtained priority is crisp and therefore does not need additional defuzzification and normalization. However, it is difficult to infer the value of $\tilde{\lambda}_{\max}$ from $\{w_j\}$. Ignoring the dependency between the dividend and divisor of Eq. (3.14) results in a simplified version indicated with FEAi.

Example 3.3 In the previous example,

$$\sum_{j=1}^{n} \tilde{a}_{1j} = (4.543, \ 6.867, \ 10.333)$$

$$\sum_{j=1}^{n} \tilde{a}_{2j} = (1.563, \ 1.761, \ 2.310)$$

$$\sum_{j=1}^{n} \tilde{a}_{3j} = (14, \ 22, \ 28)$$

$$\sum_{j=1}^{n} \tilde{a}_{4j} = (5.367, \ 9.583, \ 14.500)$$

$$\sum_{j=1}^{n} \tilde{a}_{5j} = (5.200, \ 11.333, \ 18.000)$$

After applying Theorem 3.3, the fuzzy synthetic extent of each criterion is evaluated as

$$\tilde{s}_1 = (0.067, \ 0.133, \ 0.283)$$

$$\tilde{s}_2 = (0.022, \ 0.034, \ 0.074)$$

$$\tilde{s}_3 = (0.237, \ 0.427, \ 0.627)$$

$$\tilde{s}_4 = (0.084, \ 0.186, \ 0.364)$$

$$\tilde{s}_5 = (0.086, \ 0.220, \ 0.414)$$

The results of comparing the fuzzy synthetic extents of criteria in pairs are summarized as

$$\mathbf{V} = \begin{vmatrix} - & 1 & 0.137 & 0.791 & 0.695 \\ 0.058 & - & 0 & 0 & 0 \\ 1 & 1 & - & 1 & 1 \\ 1 & 1 & 0.346 & - & 0.891 \\ 1 & 1 & 0.461 & 1 & - \end{vmatrix}$$

For example, $V(\tilde{s}_4 \geq \tilde{s}_3) = \min(1, \ 1, \ 0.346, \ 0.891) = 0.346$. As a result,

$$\{w_i\} = \{0.137, \ 0, \ 1, \ 0.346, \ 0.461\}$$

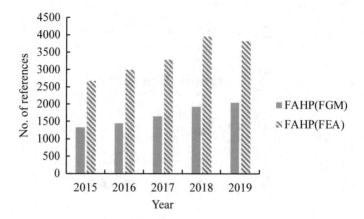

Fig. 3.2 Statistics on the number of FAHP related references (*Data source* Google Scholar)

After normalization,

$$\{w_i\} \rightarrow \{0.071,\ 0,\ 0.514,\ 0.178,\ 0.237\}$$

FGM and FEA are efficient, but approximate fuzzy priorities instead of deriving them [16]. FGM and FEA are the most prevalent techniques for deriving the priorities of criteria from pairwise comparison results [6, 17]. FEA is especially welcomed because the priorities of criteria are estimated in crisp values and do not need to be defuzzified further [18–20], as illustrated in Fig. 3.2. However, the results derived using FGM is fuzzy, which reserves much flexibility. FGM is also more similar to the original reasoning mechanism of (fuzzy) analytic hierarchy process than FEA. FGM and FEA are also subject to drawbacks such as inaccuracy, poor robustness, unreasonable priorities, information loss, and inflexibility [21].

There have been a number of variants of FEA.

3.1.4   Fuzzy Inverse of Column Sum (FICSM)

The FICSM method estimates the values of fuzzy priorities as follows [22]:

$$\tilde{w}_j \cong \frac{\frac{1}{\sum_{i=1}^{n} \tilde{a}_{ij}}}{\sum_{h=1}^{n} \frac{1}{\sum_{i=1}^{n} \tilde{a}_{ih}}} \tag{3.22}$$

Theorem 3.4 [23]

Table 3.3 Results using various methods

Method	$\tilde{w}_1$	$\tilde{w}_2$	$\tilde{w}_3$	$\tilde{w}_4$	$\tilde{w}_5$
FGM	(0.045, 0.091, 0.239)	(0.020, 0.036, 0.086)	(0.258, 0.495, 0.681)	(0.067, 0.147, 0.330)	(0.096, 0.231, 0.445)
FEA	0.071	0	0.514	0.178	0.237
FICSM	(0.040, 0.084, 0.225)	(0.023, 0.043, 0.097)	(0.284, 0.543, 0.687)	(0.057, 0.121, 0.298)	(0.099, 0.209, 0.439)

$$\tilde{w}_j \cong \left(\frac{1}{1 + \sum_{h \neq j} \frac{\sum_{i=1}^{n} a_{ij3}(k)}{\sum_{i=1}^{n} a_{ih1}(k)}}, \ \frac{1}{1 + \sum_{l \neq i} \frac{\sum_{i=1}^{n} a_{ij2}(k)}{\sum_{i=1}^{n} a_{ih2}(k)}}, \ \frac{1}{1 + \sum_{l \neq i} \frac{\sum_{i=1}^{n} a_{ij1}(k)}{\sum_{i=1}^{n} a_{ih3}(k)}} \right)$$

(3.23)

Example 3.4 FICSM is applied to Example 3.2. Then, the results using various methods are compared in Table 3.3.

3.2 Exact and Near-Exact Methods

3.2.1 Alpha Cut Operations (ACO)

Replacing the fuzzy parameters and variables in Eqs. (3.2) and (3.3) with their α cuts leads to

$$\det(\mathbf{A}(\alpha) - \lambda(\alpha)\mathbf{I}) = 0 \tag{3.24}$$

$$(\mathbf{A}(\alpha) - \lambda(\alpha)\mathbf{I})\mathbf{x}(\alpha) = 0 \tag{3.25}$$

where each α cut is an interval:

$$a_{ij}(\alpha) = [a_{ij}^L(\alpha), \ a_{ij}^R(\alpha)] \tag{3.26}$$

$$\lambda(\alpha) = [\lambda^L(\alpha), \ \lambda^R(\alpha)] \tag{3.27}$$

$$\mathbf{x}(\alpha) = [\mathbf{x^L}(\alpha), \ \mathbf{x^R}(\alpha)] \tag{3.28}$$

If α takes 11 possible values $(0, 0.1, \ldots, 1)$, Eqs. (3.24) and (3.25) must be solved $11 \cdot 2^{C_2^n}$ times to derive the membership functions of the fuzzy maximal eigenvalue and eigenvector [8]. According to the extension principle [24], from the enumeration results, the membership of each value is set to the possibly maximal membership.

Equivalently, the α cut is determined by the possibly maximal and minimal values as

$$\lambda^L(\alpha) = \min_{\det([a_{ij}^{s_{ijt}}(\alpha)] - \lambda_t(\alpha)I) = 0} (\lambda_t(\alpha)) \tag{3.29}$$

$$\lambda^R(\alpha) = \max_{\det([a_{ij}^{s_{ijt}}(\alpha)] - \lambda_t(\alpha)I) = 0} (\lambda_t(\alpha)) \tag{3.30}$$

$$\mathbf{x}^L(\alpha) = \min_{([a_{ij}^{s_{ijt}}(\alpha) - \lambda_t(\alpha)\mathbf{I}]\mathbf{x}_t(\alpha) = 0} (\mathbf{x}_t(\alpha)) \tag{3.31}$$

$$\mathbf{x}^R(\alpha) = \max_{([a_{ij}^{s_{ijt}}(\alpha) - \lambda_t(\alpha)\mathbf{I}]\mathbf{x}_t(\alpha) = 0} (\mathbf{x}_t(\alpha)) \tag{3.32}$$

where $s_{ijt} = L$ or R. $\lambda_t^L(\alpha)$, $\lambda_t^R(\alpha)$, $\mathbf{x}_t^L(\alpha)$, and $\mathbf{x}_t^R(\alpha)$ are the results derived from the t-th combination; $t = 1 \sim 11 \cdot 2^{C_2^n}$. The exact membership function of the fuzzy maximal eigenvalue is illustrated in Fig. 3.3. As expected, it is no longer a TFN. The membership function of a fuzzy priority derived using ACO is shown in Fig. 3.4.

Example 3.5 ACO is to be applied to Example 3.2. To this end, MATLAB 2017 is adopted to implement ACO. The program code is shown in Fig. 3.5.

The exact membership function of $\tilde{\lambda}_{\max}$ is shown in Fig. 3.6. The exact membership functions of the fuzzy priorities of criteria are summarized in Fig. 3.7.

Subsequently, the values of $\tilde{w}_3$ derived using various methods are compared in Fig. 3.8.

In studies, such as that of Csutora and Buckley [25], the results were derived only when $\alpha = 0$ or 1 owing to the inefficiency problem. In this way, the fuzzy maximal eigenvalue and fuzzy priorities are also represented by TFNs.

A simplified version, denoted by sACO, considers the weighted mean of α cuts instead of their individual values [26]:

Fig. 3.3 Exact membership function of the fuzzy maximal eigenvalue derived using ACO

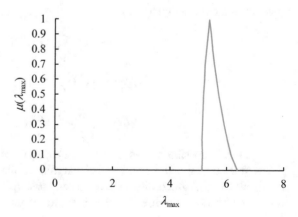

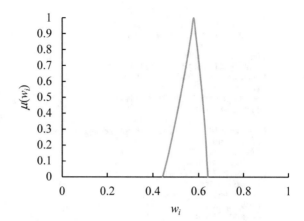

Fig. 3.4 Exact membership function of a fuzzy priority derived using ACO

$$\tilde{a}_{ij}(\alpha) \rightarrow \begin{cases} 1 & \text{if} & i = j \\ \omega a_{ij}^L(\alpha) + (1 - \omega)a_{ij}^R(\alpha) & \text{if} & a_{ij}^L(1) \geq 1 \; ; i, j = 1 \sim n \\ \dfrac{1}{\overline{a}_{ji}(\alpha)} & \text{otherwise} \end{cases}$$

(3.33)

In this way, the FAHP problem becomes a crisp AHP problem for each α value.

3.2.2 Approximating ACO (xACO) Method

In the ACO method, the α cut of the fuzzy maximal eigenvalue or a fuzzy priority is determined by the outermost points of the results after enumerating all possible combinations of $\tilde{a}_{ij}$, as illustrated in Fig. 3.9, which is a time-consuming process. To enhance the efficiency of this process, the xACO approach [8] attempts to fit the fuzzy maximal eigenvalue or a fuzzy priority with nonlinear functions during the enumeration process, based on incomplete results, as illustrated in Fig. 3.10. Thus, it is not necessary to complete the time-consuming enumeration process.

Example 3.6 In Example 3.5, if the α cuts of $\tilde{w}_1$ when $\alpha = 0, 0.5$ and 1 are known (Table 3.4). Then, the membership function of $\tilde{w}_3$ can be approximated with two logarithmic functions:

Left-hand side: $\mu_{\tilde{w}_3}(x) \cong 1.663 \ln(x) + 4.958 \; \forall \; 0.051 \leq x < 0.093$
Right-hand side: $\mu_{\tilde{w}_3}(x) \cong -1.255 \ln(x) - 1.995 \; \forall \; 0.093 \leq x < 0.206$

The approximation results are shown in Fig. 3.11. Obviously, the xACO result approximates the exact membership function very well.

When the left-hand side (or right-hand side) of the membership function of a fuzzy priority is approximated with a logarithmic function $\xi_1 \ln x_k + \zeta_1$, the following constrained optimization problem is solved:

```
stime=now       % start time
fl=zeros(11,2)        % fuzzy eigenvalue
fev=zeros(5,11,2)        % alpha cuts of fuzzy eigenvector; alpha = 0, 0.1, ..., 1

for i0=1:11
   alpha0=i0*0.1-0.1        % alpha
lmin=9999
lmax=0
evmin=[9999 9999 9999 9999 9999]
evmax=[0 0 0 0 0]

FA=zeros(5,5,3)        % fuzzy pairwise comparison matrix
FA(1,2,:)=[3 5 7]
FA(3,1,:)=[3 5 7]
FA(3,2,:)=[7 9 9]
FA(3,4,:)=[2 4 6]
FA(3,5,:)=[1 3 5]
FA(4,1,:)=[1 3 5]
FA(4,2,:)=[3 5 7]
FA(5,1,:)=[1 3 5]
FA(5,2,:)=[2 4 6]
FA(5,4,:)=[1 3 5]
FA(:,:,1)=(1-alpha0)*FA(:,:,1)+alpha0*FA(:,:,2)        % calculate alpha cuts
FA(:,:,3)=(1-alpha0)*FA(:,:,3)+alpha0*FA(:,:,2)

A=zeros(5,5)        % crisp pairwise comparison matrix
A(1,1)=1
A(2,2)=1
A(3,3)=1
A(4,4)=1
A(5,5)=1

for i1=1:2
    for i2=1:2
        for i3=1:2
            for i4=1:2
                for i5=1:2
                    for i6=1:2
                        for i7=1:2
                            for i8=1:2
                                for i9=1:2
                                    for i10=1:2

                                        % build crisp pairwise comparison matrix
                                        if FA(1,2,i1*2-1)==0
                                            A(1,2)=1/FA(2,1,i1*2-1)
                                        else
                                            A(1,2)=FA(1,2,i1*2-1)
                                        end
                                        A(2,1)=1/A(1,2)

......

                                        if FA(4,5,i10*2-1)==0
                                            A(4,5)=1/FA(5,4,i10*2-1)
                                        else
                                            A(4,5)=FA(4,5,i10*2-1)
                                        end
                                        A(5,4)=1/A(4,5)

                                        % derive the eigenvalue and eigenvector
                                        [E, V] = eig(A)
```

Fig. 3.5 MATLAB program code for implementing ACO

```
                              % update alpha cuts of fuzzy eigenvalue and eigenvector
                              if V(1,1)<lmin
                                  lmin=V(1,1)
                              end
                              if V(1,1)>lmax
                                  lmax=V(1,1)
                              end
                              ev=E(:,1)/sum(E(:,1))

                              for j=1:5
                                  if ev(j)<evmin(j)
                                      evmin(j)=ev(j)
                                  end
                              end
                              for j=1:5
                                  if ev(j)>evmax(j)
                                      evmax(j)=ev(j)
                                  end
                              end

                      end
                    end
                  end
                end
              end
            end
          end
        end
      end
    end

    % update fuzzy eigenvalue and eigenvector
    fl(i0,1)=lmin
    fl(i0,2)=lmax
    for j=1:5
      fev(j,i0,1)=evmin(j)
      fev(j,i0,2)=evmax(j)
    end

  end

  etime=now        % end time
  (etime-stime)*60*60*24        % elapsed time
```

Fig. 3.5 (continued)

$$\text{Min } Z_1 = \sum_{k=1}^{K} (\xi_1 \ln x_k + \zeta_1 - \mu_{\tilde{\lambda}_{\max}}(x_k))^2 \tag{3.34}$$

s.t.

$$\xi_1 \ln x_k + \zeta_1 \geq \mu_{\tilde{\lambda}_{\max}}(x_k) \tag{3.35}$$

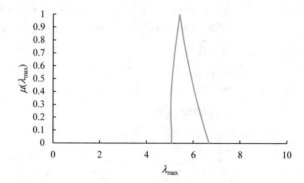

Fig. 3.6 Exact membership function of the fuzzy maximal eigenvalue in Example 3.2

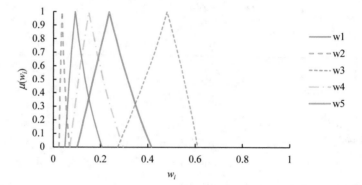

Fig. 3.7 Exact membership functions of fuzzy priorities in Example 3.2

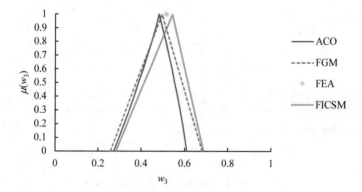

Fig. 3.8 Values of $\tilde{w}_3$ derived using various methods

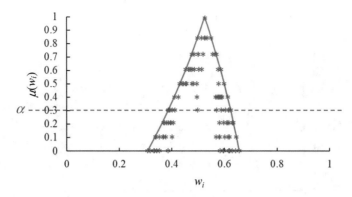

Fig. 3.9 The enumeration process of the ACO method

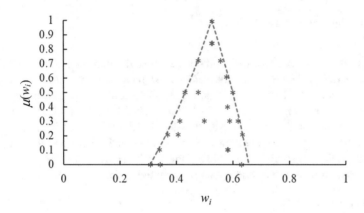

Fig. 3.10 Fitting a fuzzy priority with nonlinear functions in the process

Table 3.4 α cuts of $\tilde{w}_3$ when $\alpha = 0, 0.5$ and 1

α	$\tilde{w}_3(\alpha)$
0	[0.051, 0.206]
0.5	[0.068, 0.134]
1	[0.093, 0.093]

The constrained optimization problem can be converted into the following unconstrained optimization problem:

$$\text{Min } Z_2 = \sum_{k=1}^{K} ((\xi_1 \ln x_k + \zeta_1 - \mu_{\tilde{\lambda}_{\max}}(x_k))^2 + M(\mu_{\tilde{\lambda}_{\max}}(x_k) - \xi_1 \ln x_k - \zeta_1))$$

(3.36)

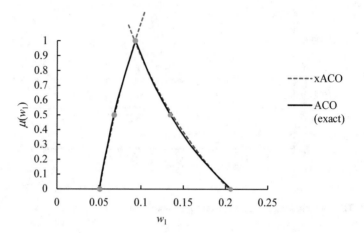

Fig. 3.11 Approximation results

where M is a large positive real number denoting the penalty for the violation of Constraint (3.35). However, as the value of M increases, the range of the fuzzy priority is widened. Therefore, a large value of M should not be set. In Eq. (3.34), the actual membership $\mu_{\tilde{\lambda}_{\max}}(x_k)$ is less than 1. Therefore, the (absolute) deviation between the estimated membership $\xi_1 \ln x_k + \zeta_1$ and the actual membership $\mu_{\tilde{\lambda}_{\max}}(x_k)$ is within $[0, 1]$, and its expectation value is less than 0.5. Therefore, it is suggested that M should be as small as possible but greater than 0.5.

The optimal values of ξ_1 and ζ_1 can be obtained as follows.

Theorem 3.5 [8]

$$\xi_1^* = \frac{K \sum_{k=1}^{K} \mu_{\tilde{\lambda}_{\max}}(x_k) \ln x_k - \sum_{k=1}^{K} \ln x_k \sum_{k=1}^{K} \mu_{\tilde{\lambda}_{\max}}(x_k)}{K \sum_{k=1}^{K} (\ln x_k)^2 - \left(\sum_{k=1}^{K} \ln x_k\right)^2} \tag{3.37}$$

$$\zeta_1 = \frac{2 \sum_{k=1}^{K} \mu_{\tilde{\lambda}_{\max}}(x_k) + KM - 2 \sum_{k=1}^{K} \ln x_k \xi_1}{2K} \tag{3.38}$$

In Fig. 3.12, the logarithmic functions were fitted using Theorem 3.5 based on 10 points generated during the enumeration process. These points are not α cuts yet. The results obtained using the ACO method are presented in this figure as well, which is based on $10 \cdot 2^{C_2^4} + 1 = 641$ points. The logarithmic functions fitted using the xACO approach are quite close to the exact membership function derived using the ACO method, but the xACO approach requires considerably less time to arrive at the results, which makes it more efficient. The lines in this figure are approximate functions, not the actual membership function. Therefore, the lines may not intersect at the core.

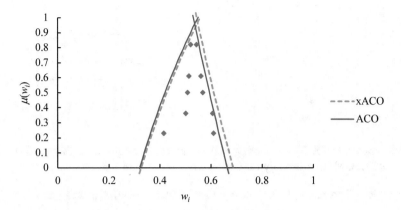

Fig. 3.12 Fitted logarithmic functions based on incomplete results

Other nonlinear function, such as polynomial functions and exponential functions, can be applied in similar manner.

3.2.3 Piecewise Linear FGM (PLFGM) Method

The required calculation in the xACO method can be simplified by taking the following treatments:

(1) Adopting piecewise linear functions to approximate the membership function of the fuzzy maximal eigenvalue or a fuzzy priority.
(2) Applying FGM instead of ACO to derive the fuzzy maximal eigenvalue and fuzzy priorities.

For this purpose, Wu et al. [3] proposed the piecewise linear FGM (PLFGM) method.

Letting the left and right α cuts of $\tilde{w}_i$ be indicated with $w_i^L(\alpha)$ and $w_i^R(\alpha)$, respectively. Then,

$$w_i^L(\alpha) \cong \frac{1}{1 + \sum_{k \neq i} \frac{\sqrt[n]{\prod_{j=1}^n a_{kj}^R(\alpha)}}{\sqrt[n]{\prod_{j=1}^n a_{ij}^L(\alpha)}}} \tag{3.39}$$

$$w_i^R(\alpha) \cong \frac{1}{1 + \sum_{k \neq i} \frac{\sqrt[n]{\prod_{j=1}^n a_{kj}^L(\alpha)}}{\sqrt[n]{\prod_{j=1}^n a_{ij}^R(\alpha)}}} \tag{3.40}$$

In PLFGM, a fuzzy priority is estimated by connecting some of its α cuts with straight lines, as illustrated in Fig. 3.13, in which the membership function on either

Fig. 3.13 Fuzzy priority
estimated using PLFGM

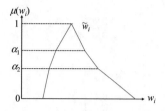

side is approximated by connecting four α cuts with straight lines. FGM is a special
case of PLFGM because only the α cuts when $\alpha = 0$ and 1 are connected.

Example 3.7 In Example 3.1, the α cuts of $\tilde{A}$ when $\alpha = 0, 0.5$, and 1 are

$$\tilde{A}(0) = \begin{vmatrix} [1,\ 1] & [2,\ 6] & [3,\ 7] \\ [1/6,\ 1/2] & [1,\ 1] & [1,\ 4] \\ [1/7,\ 1/3] & [1/4,\ 1/1] & [1,\ 1] \end{vmatrix}$$

$$\tilde{A}(0.5) = \begin{vmatrix} [1,\ 1] & [3,\ 5] & [4,\ 6] \\ [1/5,\ 1/3] & [1,\ 1] & [1.5,\ 3] \\ [1/6,\ 1/4] & [1/3,\ 1/1.5] & [1,\ 1] \end{vmatrix}$$

$$\tilde{A}(1) = \begin{vmatrix} [1,\ 1] & [4,\ 4] & [5,\ 5] \\ [1/4,\ 1/4] & [1,\ 1] & [2,\ 2] \\ [1/5,\ 1/5] & [1/2,\ 1/2] & [1,\ 1] \end{vmatrix}$$

According to Eqs. (3.39) and (3.40),

$$w_1^L(0) \cong \frac{1}{1 + \frac{\sqrt[3]{1/2 \cdot 1 \cdot 4}}{\sqrt[3]{1 \cdot 2 \cdot 3}} + \frac{\sqrt[3]{1/3 \cdot 1/1 \cdot 1}}{\sqrt[3]{1 \cdot 2 \cdot 3}}} = 0.48$$

$$w_1^R(0) \cong \frac{1}{1 + \frac{\sqrt[3]{1/6 \cdot 1 \cdot 1}}{\sqrt[3]{1 \cdot 6 \cdot 7}} + \frac{\sqrt[3]{1/7 \cdot 1/4 \cdot 1}}{\sqrt[3]{1 \cdot 6 \cdot 7}}} = 0.80$$

$$w_1^L(0.5) \cong \frac{1}{1 + \frac{\sqrt[3]{1/3 \cdot 1 \cdot 3}}{\sqrt[3]{1 \cdot 3 \cdot 4}} + \frac{\sqrt[3]{1/4 \cdot 1/1.5 \cdot 1}}{\sqrt[3]{1 \cdot 3 \cdot 4}}} = 0.60$$

$$w_1^R(0.5) \cong \frac{1}{1 + \frac{\sqrt[3]{1/5 \cdot 1 \cdot 1.5}}{\sqrt[3]{1 \cdot 5 \cdot 6}} + \frac{\sqrt[3]{1/6 \cdot 1/3 \cdot 1}}{\sqrt[3]{1 \cdot 5 \cdot 6}}} = 0.75$$

$$w_1^L(1) = w_1^R(1) \cong \frac{1}{1 + \frac{\sqrt[3]{1/4 \cdot 1 \cdot 2}}{\sqrt[3]{1 \cdot 4 \cdot 5}} + \frac{\sqrt[3]{1/5 \cdot 1/2 \cdot 1}}{\sqrt[3]{1 \cdot 4 \cdot 5}}} = 0.68$$

As a result, the piecewise linear membership function of $\tilde{w}_1$ is

Fig. 3.14 Fitted piecewise linear membership function

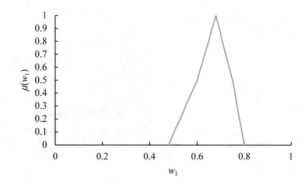

$$\mu_{\tilde{w}_1}(x) \cong \begin{cases} 0 & if & x < 0.48 \\ \frac{0.5-0}{0.60-0.48}(x-0.48)+0 & if & 0.48 \le x < 0.60 \\ \frac{1-0.5}{0.68-0.60}(x-0.60)+0.5 & if & 0.60 \le x < 0.68 \\ \frac{0.5-1}{0.75-0.68}(x-0.68)+1 & if & 0.68 \le x < 0.75 \\ \frac{0-0.5}{0.8-0.75}(x-0.75)+0.5 & if & 0.75 \le x < 0.8 \\ 0 & otherwise \end{cases}$$

which is illustrated in Fig. 3.14.

3.3   Fuzzy Measuring Attractiveness by a Categorical Based Evaluation Technique (Fuzzy MACBETH)

Fuzzy MACBETH is also based on pairwise comparison results. However, fuzzy MACBETH uses an interval scale, rather than a ratio scale. In addition, the calculation process of fuzzy MACBETH is also different [27].

In fuzzy MACBETH, the relative priority of a criterion over another, or the relative attractiveness of an alternative over another, is classified into semantic categories including "not different from", "very weakly more important (or attractive) than", "weakly more important (or attractive) than", "moderately more important (or attractive) than", "strongly more important (or attractive) than", "very strongly more important (or attractive) than" and "extremely more important (or attractive) than". These semantic terms are usually mapped to TFNs within [1, 7], as illustrated in Table 3.5.

Based on pairwise comparison results, a fuzzy judgment matrix $\tilde{\mathbf{A}}$ is constructed. More important critical features will be placed closer to the upper left corner. Therefore,

$$\tilde{w}_i \ge \tilde{w}_j \; \forall i < j \tag{3.41}$$

Table 3.5 Semantic terms for pairwise comparison

Semantic term	TFN
Not different from	(1, 1, 3)
Very weekly more important than	(1, 2, 4)
Weakly more important than	(1, 3, 5)
Moderately more important than	(2, 4, 6)
Strongly more important than	(3, 5, 7)
Very strongly more important than	(4, 6, 7)
Extremely more important than	(5, 7, 7)

There are various ways to derive the fuzzy priorities of criteria from $\tilde{\mathbf{A}}$ [27]. For example, the following fuzzy quadratic programming (FQP) model can be optimized to fulfill this purpose:

(FQP Model)

$$\text{Min } \tilde{Z} = \sum_{i=1}^{n} \sum_{j \neq i} (\tilde{w}_i(-)\tilde{w}_j(-)(\tilde{a}_{ij} - 1)c)^2 \tag{3.42}$$

subject to

$$\tilde{w}_1 = 100 \tag{3.43}$$

$$\tilde{w}_n = 0 \tag{3.44}$$

$$0 \leq \tilde{w}_i \leq 100 \; \forall i \in [2, n-1] \tag{3.45}$$

$$c \geq 0 \tag{3.46}$$

The FQP problem needs to be converted into a crisp problem to be easily solved. First, according to the arithmetic for TFNs [28], the objective function is decomposed into

$$Z_1 = \sum_{i=1}^{n} \sum_{j \neq i} (w_{i1} - w_{j3} - a_{ij3}c + c)^2 \tag{3.47}$$

$$Z_2 = \sum_{i=1}^{n} \sum_{j \neq i} (w_{i2} - w_{j2} - a_{ij2}c + c)^2 \tag{3.48}$$

$$Z_3 = \sum_{i=1}^{n} \sum_{j \neq i} (w_{i3} - w_{j1} - a_{ij1}c + c)^2 \tag{3.49}$$

A common treatment is to defuzzify $\tilde{Z}$ using COG [28] as

$$D(\tilde{Z}) = \frac{Z_1 + Z_2 + Z_3}{3} \tag{3.50}$$

In addition, Constraints (3.43) and (3.44) are fuzzily equivalent to

$$w_{11} = 50 \tag{3.51}$$

$$w_{12} = w_{13} = 100 \tag{3.52}$$

$$w_{n1} = w_{n2} = 0 \tag{3.53}$$

$$w_{n3} = 50 \tag{3.54}$$

Constraint (3.45) can be decomposed into

$$w_{i1} \geq 0 \tag{3.55}$$

$$w_{i3} \leq 100 \tag{3.56}$$

Finally, the following quadratic programming problem is solved instead:

(QP Model)

$$\text{Min } D(\tilde{Z}) = \frac{Z_1 + Z_2 + Z_3}{3} \tag{3.57}$$

subject to

$$Z_1 = \sum_{i=1}^{n} \sum_{j \neq i} (w_{i1} - w_{j3} - a_{ij3}c + c)^2 \tag{3.58}$$

$$Z_2 = \sum_{i=1}^{n} \sum_{j \neq i} (w_{i2} - w_{j2} - a_{ij2}c + c)^2 \tag{3.59}$$

$$Z_3 = \sum_{i=1}^{n} \sum_{j \neq i} (w_{i3} - w_{j1} - a_{ij1}c + c)^2 \tag{3.60}$$

$$w_{11} = 50 \tag{3.61}$$

$$w_{12} = w_{13} = 100 \tag{3.62}$$

$$w_{n1} = w_{n2} = 0 \tag{3.63}$$

$$w_{n3} = 50 \tag{3.64}$$

$$w_{i1} \geq 0 \ \forall i \in [2, n-1] \tag{3.65}$$

$$w_{i3} \leq 100 \ \forall i \in [2, n-1] \tag{3.66}$$

$$c \geq 0 \tag{3.67}$$

$$w_{i1} \leq w_{i2} \leq w_{i3} \ \forall i \tag{3.68}$$

Pairwise comparison results are inconsistent if $D(\tilde{Z}) \geq \xi$ where ξ is a threshold. The optimal solution needs to be normalized:

$$
\begin{aligned}
\tilde{w}_i &\rightarrow \frac{\tilde{w}_i}{\sum_{j=1}^{n} \tilde{w}_j} \\
&= \frac{1}{1 + \sum_{j \neq i} \frac{\tilde{w}_j}{\tilde{w}_i}}
\end{aligned}
\tag{3.69}
$$

except that $w_{n1} = w_{n2} = 0$. Fuzzy priorities derived by solving QP Model may be symmetric. Nevertheless, after normalization, fuzzy priorities become asymmetric.

Example 3.8 Fuzzy MACBETH is applied to derive the fuzzy priorities of criteria from the following fuzzy judgment matrix:

$$
\tilde{A} =
\begin{vmatrix}
1 & (1, \ 3, \ 5) & (3, \ 5, \ 7) & (2, \ 4, \ 6) \\
- & 1 & (1, \ 3, \ 5) & (3, \ 5, \ 7) \\
- & - & 1 & (2, \ 4, \ 6) \\
- & - & - & 1
\end{vmatrix}
$$

Lingo is applied to encode the QP Model, as shown in Fig. 3.15. The derived fuzzy priorities of criteria are summarized in Fig. 3.16.

```
min=1/3*(Z1+Z2+Z3);
Z1=(w11-w23-5*c+c)^2+(w11-w33-6*c+c)^2+(w11-w43-7*c+c)^2+(w11-w53-7*c+c)^2+(w21-
w33-4*c+c)^2+(w21-w43-7*c+c)^2+(w21-w53-7*c+c)^2+(w31-w43-7*c+c)^2+(w31-w53-7*c+c
)^2+(w41-w53-3*c+c)^2;
Z2=(w12-w22-3*c+c)^2+(w12-w32-2*c+c)^2+(w12-w42-5*c+c)^2+(w12-w52-7*c+c)^2+(w22-
w32-1*c+c)^2+(w22-w42-5*c+c)^2+(w22-w52-5*c+c)^2+(w32-w42-5*c+c)^2+(w32-w52-6*c+c
)^2+(w42-w52-1*c+c)^2;
Z3=(w13-w21-1*c+c)^2+(w13-w31-2*c+c)^2+(w13-w41-3*c+c)^2+(w13-w51-5*c+c)^2+(w23-
w31-1*c+c)^2+(w23-w41-3*c+c)^2+(w23-w51-3*c+c)^2+(w33-w41-3*c+c)^2+(w33-w51-4*c+c
)^2+(w43-w51-1*c+c)^2;
w11=50;
w12=100;
w13=100;
w51=0;
w52=0;
w53=50;
w21>=0;
w31>=0;
w41>=0;
w23<=100;
w33<=100;
w43<=100;
w21<=w22;
w22<=w23;
w31<=w32;
w32<=w33;
w41<=w42;
w42<=w43;
```

Fig. 3.15   QP Model

Fig. 3.16   Derived fuzzy
priorities of criteria

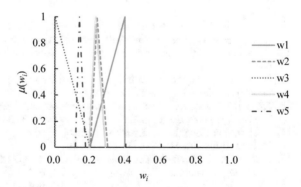

References

1. D. Choudhary, R. Shankar, An STEEP-fuzzy AHP-TOPSIS framework for evaluation and selection of thermal power plant location: a case study from India. Energy **42**(1), 510–521 (2012)
2. S.K. Patil, R. Kant, A fuzzy AHP-TOPSIS framework for ranking the solutions of knowledge management adoption in supply chain to overcome its barriers. Expert Syst. Appl. **41**(2), 679–693 (2014)
3. H.-C. Wu, T. Chen, C.-H. Huang, A piecewise linear FGM approach for efficient and accurate FAHP analysis: smart backpack design as an example. Mathematics **8**, 1319 (2020)
4. S.H. Zyoud, L.G. Kaufmann, H. Shaheen, S. Samhan, D. Fuchs-Hanusch, A framework for water loss management in developing countries under fuzzy environment: integration of Fuzzy AHP with Fuzzy TOPSIS. Expert Syst. Appl. **61**, 86–105 (2016)
5. H.-C. Wu, T.-C.T. Chen, C.-H. Huang, Y.-C. Shi, Comparing built-in power banks for a smart backpack design using an auto-weighting fuzzy-weighted-intersection FAHP approach. Mathematics **8**(10), 1759 (2020)
6. G. Zheng, N. Zhu, Z. Tian, Y. Chen, B. Sun, Application of a trapezoidal fuzzy AHP method for work safety evaluation and early warning rating of hot and humid environments. Saf. Sci. **50**(2), 228–239 (2012)
7. N.G. Seresht, A.R. Fayek, Computational method for fuzzy arithmetic operations on triangular fuzzy numbers by extension principle. Int. J. Approx. Reason. **106**, 172–193 (2019)
8. T. Chen, Y.C. Lin, M.C. Chiu, Approximating alpha-cut operations approach for effective and efficient fuzzy analytic hierarchy process analysis. Appl. Soft Comput. **85**, 105855 (2019)
9. J.J. Buckley, Fuzzy hierarchical analysis. Fuzzy Sets Syst. **17**(3), 233–247 (1985)
10. T. Chen, A FAHP-FTOPSIS approach for choosing mid-term occupational healthcare measures amid the COVID-19 pandemic. Health Policy Technol. **10**(2), 100517 (2021)
11. T.L. Saaty, Decision making with the analytic hierarchy process. Int. J. Serv. Sci. **1**(1), 83–98 (2008)
12. T. Chen, Y.-C. Wang, A calibrated piecewise-linear FGM approach for travel destination recommendation amid the COVID-19 pandemic. Appl. Soft Comput. **109**, 107535 (2021)
13. D.Y. Chang, Applications of the extent analysis method on fuzzy AHP. Eur. J. Oper. Res. **95**, 649–655 (1996)
14. T. Chen, Evaluating the sustainability of a smart technology application to mobile health care—the FGM-ACO-FWA approach. Complex Intell. Syst. **6**, 109–121 (2020)
15. F. Liu, J.M. Mendel, Aggregation using the fuzzy weighted average as computed by the Karnik-Mendel algorithms. IEEE Trans. Fuzzy Syst. **16**(1), 1–12 (2008)
16. T. Chen, Enhancing the efficiency and accuracy of existing FAHP decision-making methods. EURO J. Decis. Process. **8**(3), 177–204 (2020)
17. P. Sirisawat, T. Kiatcharoenpol, Fuzzy AHP-TOPSIS approaches to prioritizing solutions for reverse logistics barriers. Comput. Ind. Eng. **117**, 303–318 (2018)
18. F.R.L. Junior, L. Osiro, L.C.R. Carpinetti, A comparison between Fuzzy AHP and Fuzzy TOPSIS methods to supplier selection. Appl. Soft Comput. **21**, 194–209 (2014)
19. A. Güran, M. Uysal, Y. Ekinci, C.B. Güran, An additive FAHP based sentence score function for text summarization. Inf. Technol. Control **46**(1), 53–69 (2017)
20. A. Gnanavelbabu, P. Arunagiri, Ranking of MUDA using AHP and Fuzzy AHP algorithm. Mater. Today: Proc. **5**(5–2), 13406–13412 (2018)
21. S. Kubler, J. Robert, W. Derigent, A. Voisin, Y. Le Traon, A state-of the-art survey & testbed of fuzzy AHP (FAHP) applications. Expert Syst. Appl. **65**, 398–422 (2016)
22. F. Ahmed, K. Kilic, Fuzzy analytic hierarchy process: a performance analysis of various algorithms. Fuzzy Sets Syst. **362**, 110–128 (2019)
23. M.-C. Chiu, T. Chen, Assessing mobile and smart technology applications to active and healthy ageing using a fuzzy collaborative intelligence approach. Cognit. Comput. **13**(2), 431–446 (2021)

24. L.A. Zadeh, Fuzzy logic = computing with words. IEEE Trans. Fuzzy Syst. **4**(2), 103–111 (1996)
25. R. Csutora, J.J. Buckley, Fuzzy hierarchical analysis: the Lambda-Max method. Fuzzy Sets Syst. **120**(2), 181–195 (2001)
26. M.A.B. Promentilla, T. Furuichi, K. Ishii, N. Tanikawa, A fuzzy analytic network process for multi-criteria evaluation of contaminated site remedial countermeasures. J. Environ. Manag. **88**(3), 479–495 (2008)
27. D. Dhouib, An extension of MACBETH method for a fuzzy environment to analyze alternatives in reverse logistics for automobile tire wastes. Omega **42**(1), 25–32 (2014)
28. M. Hanss, *Applied Fuzzy Arithmetic* (Springer, Berlin Heidelberg, 2005)

Chapter 4
Consensus Measurement and Enhancement

4.1 Similarity-Based and Proximity-Based Consensus Measurement

4.1.1 Consensus on Judgments

The consensus among decision makers is usually measured by the similarity or proximity of their judgments (or decisions) [1–3]. For example, in fuzzy analytic hierarchy process (FAHP) methods [4], assume that the fuzzy judgment matrix of decision maker k is indicated by $\tilde{\mathbf{A}}(k) = [\tilde{a}_{ij}(k)]$. Then, the similarity between the fuzzy judgment matrixes of two decision makers k and l can be measured by their concordance [1]:

$$sim(\tilde{\mathbf{A}}(k), \ \tilde{\mathbf{A}}(l)) = \frac{\sum_{i=1}^{n-1} \sum_{j=i+1}^{n} I(\tilde{a}_{ij}(k), \ \tilde{a}_{ij}(l))}{0.5n(n-1)} \qquad (4.1)$$

where

$$I(\tilde{a}_{ij}(k), \ \tilde{a}_{ij}(l)) = \begin{cases} 1 & if \quad \tilde{a}_{ij}(k) = \tilde{a}_{ij}(l) \\ 0 & otherwise \end{cases} \qquad (4.2)$$

Then,

(1) If $sim(\tilde{\mathbf{A}}(k), \ \tilde{\mathbf{A}}(l))$ is higher than a threshold, the two decision makers have reached a consensus on the relative priorities of criteria.
(2) If every pair of decision makers reaches a consensus, then the overall consensus among all decision makers exists.

However, $\tilde{a}_{ij}(k)$ is seldom the same as $\tilde{a}_{ij}(l)$. Nevertheless, they may overlap, which is utilized to measure their similarity, as illustrated in Fig. 4.1.

T.-C. T. Chen, *Advances in Fuzzy Group Decision Making*, SpringerBriefs in Applied Sciences and Technology, https://doi.org/10.1007/978-3-030-86208-4_4

Fig. 4.1 Utilizing the overlapping area of two judgments to measure their similarity

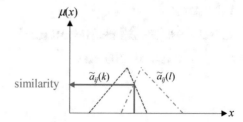

$$sim(\tilde{\mathbf{A}}(k),\ \tilde{\mathbf{A}}(l)) = \frac{\sum_{i=1}^{n-1}\sum_{j=i+1}^{n}\sup_{x}\min(\mu_{\tilde{a}_{ij}(k)}(x),\ \mu_{\tilde{a}_{ij}(l)}(x))}{0.5n(n-1)} \tag{4.3}$$

$0 \leq sim(\tilde{\mathbf{A}}(k),\ \tilde{\mathbf{A}}(l)) \leq 1$.

The proximity between two fuzzy judgment matrixes can be measured in terms of their fuzzy Frobenius distance [5]:

$$\widetilde{dist}(\tilde{\mathbf{A}}(k),\ \tilde{\mathbf{A}}(l)) = \sqrt{\sum_{i=1}^{n}\sum_{j=1}^{n}(\tilde{a}_{ij}(k)(-)\tilde{a}_{ij}(l))^2} \tag{4.4}$$

$(-)$ denotes fuzzy subtraction. However, a_{ij1} and a_{ij3} are usually dependent on a_{ij2}. For example, in Zheng et al. [6],

$$a_{ij1}(k) = \max(1,\ a_{ij2}(k) - 2) \tag{4.5}$$

$$a_{ij3}(k) = \min(9,\ a_{ij2}(k) + 2) \tag{4.6}$$

Therefore, calculating the crisp Frobenius distance is sufficient [7]:

$$dist(\tilde{\mathbf{A}}(k),\ \tilde{\mathbf{A}}(l)) = \sqrt{\sum_{i=1}^{n}\sum_{j=1}^{n}(a_{ij2}(k) - a_{ij2}(l))^2} \tag{4.7}$$

In contrast, some studies [8, 9] adopted the following distance measure:

$$dist(\tilde{\mathbf{A}}(k),\ \tilde{\mathbf{A}}(l)) = \sum_{i=1}^{n}\sum_{j=1}^{n}\frac{|a_{ij1}(k) - a_{ij1}(l)| + |a_{ij2}(k) - a_{ij2}(l)| + |a_{ij3}(k) - a_{ij3}(l)|}{3} \tag{4.8}$$

There are also studies [10] that assigned unequal weights to different pairwise comparison results:

$$dist(\tilde{\mathbf{A}}(k),\ \tilde{\mathbf{A}}(l)) = \sqrt{\sum_{i=1}^{n}\sum_{j=1}^{n}v_{ij}(a_{ij2}(k) - a_{ij2}(l))^2} \tag{4.9}$$

where υ_{ij} is the weight assigned to $\tilde{a}_{ij}$.

Then, the proximity between the two fuzzy judgment matrixes can be measured as

$$prox(\tilde{\mathbf{A}}(k), \ \tilde{\mathbf{A}}(l)) = e^{-dist(\tilde{\mathbf{A}}(k), \ \tilde{\mathbf{A}}(l))} \tag{4.10}$$

$0 \leq prox(\tilde{\mathbf{A}}(k), \ \tilde{\mathbf{A}}(l)) \leq 1$. According to the measurement result,

(1) If $prox(\tilde{\mathbf{A}}(k), \ \tilde{\mathbf{A}}(l))$ is higher than a threshold, the two decision makers have reached a consensus on the relative priorities of criteria.
(2) If every pair of decision makers reaches a consensus, then the overall consensus among all decision makers exists.

However, in these ways, the fuzzy judgment matrixes of decision makers need to be compared in pairs, which is a time-consuming task. As an alternative, the fuzzy judgment matrixes of decision makers are averaged:

$$\overline{\tilde{\mathbf{A}}} = \frac{\sum_{k=1}^{K} \tilde{\mathbf{A}}(k)}{K} \tag{4.11}$$

Then, the fuzzy judgment matrix of each decision maker is compared with the average, which reduces the number of comparisons from $0.5K(K-1)$ to K.

4.1.2 Consensus on the Fuzzy Priorities of Criteria

The number of criteria is much fewer than the number of pairwise comparison results. Therefore, comparing the fuzzy priorities of criteria derived by decision makers is easier. For this purpose, a similarity-based consensus measure is

$$sim(\tilde{\mathbf{w}}(k), \ \tilde{\mathbf{w}}(l)) = \frac{\sum_{i=1}^{n} \sup_{x} \min(\mu_{\tilde{w}_i(k)}(x), \ \mu_{\tilde{w}_i(l)}(x))}{n} \tag{4.12}$$

where $\tilde{\mathbf{w}}(k)$ is the set of the fuzzy priorities of criteria derived by decision maker k. An example is given below.

Example 4.1 The fuzzy priorities of three criteria derived by two decision makers are summarized in Table 4.1. After applying Eq. (4.12) to these data,

(1) The similarity between the values of $\tilde{w}_1$ derived by the two decision makers is 0.75.
(2) The similarity between the values of $\tilde{w}_2$ derived by the two decision makers is 0.79.
(3) The similarity between the values of $\tilde{w}_3$ derived by the two decision makers is 0.54.

Table 4.1 Priorities of criteria derived by two decision makers

Decision maker	$\tilde{w}_1$	$\tilde{w}_2$	$\tilde{w}_3$
#1	(0.16, 0.24, 0.33)	(0.39, 0.53, 0.70)	(0.09, 0.23, 0.41)
#2	(0.11, 0.19, 0.31)	(0.33, 0.46, 0.66)	(0.27, 0.35, 0.47)

Therefore, the similarity between the fuzzy priorities of criteria derived by the two decision makers is $(0.75 + 0.79 + 0.54)/3 = 0.69$. Assume that a threshold of 0.65 is established. Obviously, $0.69 > 0.65$, showing the two decision makers have reached a consensus on the fuzzy priorities of criteria.

The proximity between the fuzzy priorities of criteria derived by two decision makers can also be measured as

$$\widetilde{prox}(\tilde{\mathbf{w}}(k),\ \tilde{\mathbf{w}}(l)) = e^{-\widetilde{dist}(\tilde{\mathbf{w}}(k),\ \tilde{\mathbf{w}}(l))} \tag{4.13}$$

where

$$\widetilde{dist}(\tilde{\mathbf{w}}(k),\ \tilde{\mathbf{w}}(l)) = \sqrt{\sum_{i=1}^{n}(\tilde{w}_i(k)(-)\tilde{w}_i(l))^2} \tag{4.14}$$

Theorem 4.1

$$\widetilde{dist}(\tilde{\mathbf{w}}(k),\ \tilde{\mathbf{w}}(l)) \cong \left(\sqrt{\sum_{i=1}^{n}\min(\max(w_{i1}(k) - w_{i3}(l),\ 0),\ \max(w_{i1}(l) - w_{i3}(k),\ 0))^2}\right.$$

$$\sqrt{\sum_{i=1}^{n}(w_{i2}(k) - w_{i2}(l))^2}$$

$$\left.\sqrt{\sum_{i=1}^{n}\max(\max(w_{i3}(k) - w_{i1}(l),\ 0),\ \max(w_{i3}(l) - w_{i1}(k),\ 0))^2}\right) \tag{4.15}$$

Theorem 4.2

$$\widetilde{prox}(\tilde{\mathbf{w}}(k),\ \tilde{\mathbf{w}}(l)) \cong (e^{-dist_3(\tilde{\mathbf{w}}(k),\ \tilde{\mathbf{w}}(l))},\ e^{-dist_2(\tilde{\mathbf{w}}(k),\ \tilde{\mathbf{w}}(l))},\ e^{-dist_1(\tilde{\mathbf{w}}(k),\ \tilde{\mathbf{w}}(l))}) \tag{4.16}$$

In the previous example,

$$\widetilde{dist}(\tilde{\mathbf{w}}(k),\ \tilde{\mathbf{w}}(l)) \cong \left(\sqrt{\begin{array}{l}\min(\max(0.16 - 0.31,\ 0),\ \max(0.11 - 0.33,\ 0))^2 \\ + \min(\max(0.39 - 0.66,\ 0),\ \max(0.33 - 0.70,\ 0))^2 \\ + \min(\max(0.09 - 0.47,\ 0),\ \max(0.27 - 0.41,\ 0))^2\end{array}}\right.$$

$$\sqrt{(0.24 - 0.19)^2 + (0.53 - 0.46)^2 + (0.23 - 0.35)^2}$$

$$\sqrt{\begin{array}{l} \max(\max(0.33-0.11,\ 0),\ \max(0.31-0.16,\ 0))^2 \\ +\max(\max(0.70-0.33,\ 0),\ \max(0.66-0.39,\ 0))^2) \\ +\max(\max(0.41-0.27,\ 0),\ \max(0.47-0.09,\ 0))^2 \end{array}}$$
$$= (0,\ 0.15,\ 0.57)$$

Therefore,

$$\widetilde{prox}(\tilde{\mathbf{w}}(1),\ \tilde{\mathbf{w}}(2)) \cong (e^{-0.57},\ e^{-0.15},\ e^{-0}) = (0.56,\ 0.86,\ 1)$$

which is high, showing a consensus between the two decision makers.

4.1.3   Consensus on the Overall Performances of Alternatives

Whether there is a consensus between two decision makers can also be evaluated by comparing the overall performances of alternatives evaluated by them:

$$sim(\tilde{\mathbf{O}}(k),\ \tilde{\mathbf{O}}(l)) = \frac{\sum_{q=1}^{Q} \sup\limits_{x} \min(\mu_{\tilde{O}_q(k)}(x),\ \mu_{\tilde{O}_q(l)}(x))}{Q} \qquad (4.17)$$

where $\tilde{\mathbf{O}}(k)$ is the set of the overall performances of alternatives evaluated by decision maker k. A proximity-based consensus measure is

$$\widetilde{prox}(\tilde{\mathbf{O}}(k),\ \tilde{\mathbf{O}}(l)) = e^{-\frac{\widetilde{dist}(\tilde{\mathbf{O}}(k),\ \tilde{\mathbf{O}}(l))}{\lambda}} \qquad (4.18)$$

where λ is a constant to rescale the fuzzy distance into a narrower range:

$$\widetilde{dist}(\tilde{\mathbf{O}}(k),\ \tilde{\mathbf{O}}(l)) = \sqrt{\sum_{q=1}^{Q} (\tilde{O}_q(k)(-)\tilde{O}_q(l))^2} \qquad (4.19)$$

The value of λ can be set to $\max\limits_{q} \max\limits_{k} O_{q3}(k)$.

Theorem 4.3

$$\widetilde{dist}(\tilde{\mathbf{O}}(k),\ \tilde{\mathbf{O}}(l)) \cong \left(\sqrt{\sum_{q=1}^{Q} \min(\max(O_{q1}(k)-O_{q3}(l),\ 0),\ \max(O_{q1}(l)-O_{q3}(k),\ 0))^2} \right.$$

$$\sqrt{\sum_{q=1}^{Q} (O_{q2}(k)-O_{q2}(l))^2}$$

$$\sqrt{\sum_{q=1}^{Q} \max(\max(O_{q3}(k)-O_{q1}(l),\ 0),\ \max(O_{q3}(l)-O_{q1}(k),\ 0))^2)} \qquad (4.20)$$

Theorem 4.4

$$\widehat{prox}(\tilde{\mathbf{O}}(k),\ \tilde{\mathbf{O}}(l)) \cong (e^{-\frac{dist_3(\tilde{\mathbf{O}}(k),\ \tilde{\mathbf{O}}(l))}{\lambda}},\ e^{-\frac{dist_2(\tilde{\mathbf{O}}(k),\ \tilde{\mathbf{O}}(l))}{\lambda}},\ e^{-\frac{dist_1(\tilde{\mathbf{O}}(k),\ \tilde{\mathbf{O}}(l))}{\lambda}}) \qquad (4.21)$$

4.2 Fuzzy Intersection-Based Consensus Measurement

4.2.1 Consensus on Judgments

The fuzzy intersection (FI) of the pairwise comparison results of decision makers, $\widetilde{FI}(\{\tilde{a}_{ij}(k)\})$, has been utilized to measure the consensus among them before deriving the criteria of priorities [11–13]:

$$\mu_{\widetilde{FI}(\{\tilde{a}_{ij}(k)\})}(x) = \min_{k}(\{\mu_{\tilde{a}_{ij}(k)}(x)\}) \qquad (4.22)$$

The FI result is the overlapping area of the membership functions, as illustrated in Fig. 4.2. Only values that are considered highly possible by all decision makers will have high memberships in the FI result.

The larger the overlapping area is, the higher is the overall consensus reached. In contrast, in Fig. 4.1, only the height of the overlapping area is concerned. If $\widetilde{FI}(\{\tilde{a}_{ij}(k)\}) = \emptyset$, the overall consensus does not exist.

Example 4.2 Three decision makers compare the relative priority of a criterion over another with the following linguistic terms:

Decision maker #1: "Weakly more important than" → (1, 3, 5)
Decision maker #2: "Weakly or strongly more important than" → (2, 4, 6)

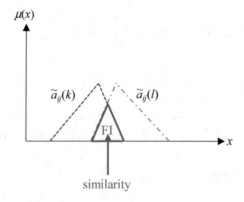

Fig. 4.2 Concept of FI

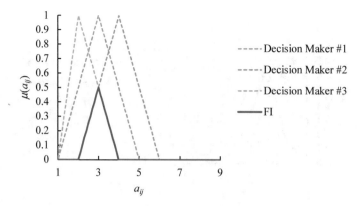

Fig. 4.3 FI of decision maker's pairwise comparison results

Decision maker #3: "As equal as or weakly more important than" → (1, 2, 4)

The FI of their pairwise comparison results is shown in Fig. 4.3. For example, the membership of 3 in the FI result is min(1, 0.5, 0.5) = 0.5. Obviously, the three decision makers have a consensus on the value of the relative priority.

Theorem 4.5 $\widetilde{FI}(\tilde{a}_{ij}(k), \tilde{a}_{ij}(l)) = \emptyset$ *if* $|a_{ij2}(k) - a_{ij2}(l)| \geq 4$ *when* $\tilde{a}_{ij}$ *is defined according to* Eqs. (4.5) *and* (4.6).

If decision makers have unequal authority levels (or weights), the fuzzy weighted intersection (FWI) operator [14, 15] *is applicable.*

Definition 4.1 The FWI of the relative priorities of criterion i over criterion j compared by K decision makers, indicated by $\tilde{a}_{ij}(1) \sim \tilde{a}_{ij}(K)$, is denoted by $\widetilde{FWI}(\tilde{a}_{ij}(1), \ldots, \tilde{a}_{ij}(K))$ such that

$$\mu_{\widetilde{FWI}}(x) = \min_l \mu_{\tilde{a}_{ij}(l)}(x) + \sum_m (\omega_m - \min_l \omega_l)(\mu_{\tilde{a}_{ij}(m)}(x) - \min_l \mu_{\tilde{a}_{ij}(l)}(x))$$

(4.23)

where ω_m is the authority level (or weight) of decision maker m; $\omega_{m_1} \neq \omega_{m_2} \exists m_1 \neq m_2$; $\sum_m \omega_m = 1$.

In the previous example, assume that the authority levels of decision makers are 0.5, 0.3, and 0.2, respectively. Then the FWI result is shown in Fig. 4.4. For example, the membership of 3 in the FWI result is min(1, 0.5, 0.5) + (0.5 − 0.2) · (1 − 0.5) + (0.3 − 0.2) · (0.5 − 0.5) + (0.2 − 0.2) · (0.5 − 0.5) = 0.65.

Only values that are considered highly possible by the most authoritative decision maker or all decision makers will have high memberships in the FWI result. If $\widetilde{FWI}(\{\tilde{a}_{ij}(k)\}) = \emptyset$, the overall consensus does not exist.

FI is a special case of FWI. Therefore, if the FI result is not an empty set, neither is the FWI result. On the contrary, it is possible that the FI result is an empty set, but

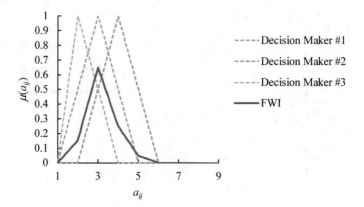

Fig. 4.4 FWI of decision maker's pairwise comparison results

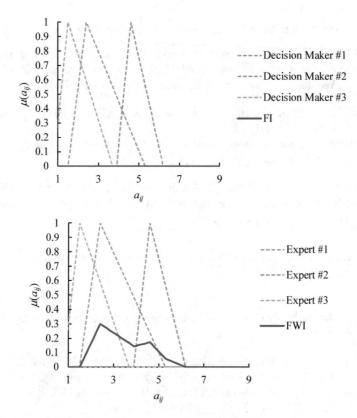

Fig. 4.5 An example that FI result is an empty set but FWI result is not

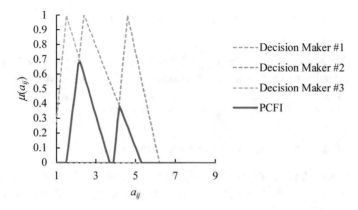

Fig. 4.6 PCFI result of the example in Fig. 4.5

the FWI result is not, as illustrated in Fig. 4.5. In this example, the decision maker whose judgment is far from the average has a low authority level.

When the overall consensus among all decision makers does not exist, the partial consensus FI (PCFI) among most decision makers [16–18] can be sought instead.

Definition 4.2 The H/K partial consensus fuzzy intersection (PCFI) of the judgments of K decision makers for the relative priority of the ith criterion over the jth criterion, indicated by $\tilde{a}_{ij}(1) \sim \tilde{a}_{ij}(K)$, is denoted by $\widetilde{PCFI}^{H/K}(\{\tilde{a}_{ij}(k)\})$ such that

$$\mu_{\widetilde{PCFI}^{H/K}}(x) = \max_g (\min(\mu_{\tilde{a}_{ij}(g(1))}(x), \ ..., \ \mu_{\tilde{a}_{ij}(g(H))}(x))) \tag{4.24}$$

where $g() \in Z^+$; $1 \le g() \le K$; $g(p) \cap g(q) = \emptyset \ \forall \ p \ne q$; $H \ge 2$.

If the H/K PCFI result is an empty set, there is a lack of partial consensus among decision makers. Therefore, the number of decision makers that reach a partial consensus needs to be reduced further.

For example, in Fig. 4.5, the three decision makers lack an overall consensus. Therefore, PCFI is applied to seek the maximal partial consensus between only two decision makers. The result is shown in Fig. 4.6. For example, the membership of 3 in the PCFI result is max(min(0.79, 0), min(0.79, 0.32), min(0, 0.32)) = 0.32, which means this value has a maximal membership of 0.32 in the membership functions of two decision makers (i.e., Decision Makers #1 and #3).

4.2.2   Consensus on the Fuzzy Priorities of Criteria

All measures proposed in the previous section can be applied to check the existence of a consensus among decision makers on the fuzzy priorities of criteria. It will be easier to reach a consensus since the number of criteria is much less than the number of pairwise comparison results.

Definition 4.3 The FI of the fuzzy priorities of criterion i derived by K decision makers, indicated by $\tilde{w}_i(1) \sim \tilde{w}_i(K)$, is denoted by $\widetilde{FI}(\{\tilde{w}_i(k)\})$ such that

$$\mu_{\widetilde{FI}(\{\tilde{w}_i(k)\})}(x) = \min_k(\{\mu_{\tilde{w}_i(k)}(x)\}) \tag{4.25}$$

If $\widetilde{FI}(\{\tilde{w}_i(k)\}) = \varnothing$, the overall consensus does not exist.

Definition 4.4 The FWI of the fuzzy priorities of criterion i derived by K decision makers, indicated by $\tilde{w}_i(1) \sim \tilde{w}_i(K)$, is denoted by $\widetilde{FWI}(\{\tilde{w}_i(k)\})$ such that

$$\mu_{\widetilde{FWI}}(x) = \min_l \mu_{\tilde{w}_i(l)}(x) + \sum_m (\omega_m - \min_l \omega_l)(\mu_{\tilde{w}_i(m)}(x) - \min_l \mu_{\tilde{w}_i(l)}(x)) \tag{4.26}$$

where ω_m is the authority level (or weight) of decision maker m; $\omega_{m_1} \neq \omega_{m_2} \exists m_1 \neq m_2$; $\sum_m \omega_m = 1$.

If $\widetilde{FWI}(\{\tilde{w}_i(k)\}) = \varnothing$, the overall consensus does not exist.

Definition 4.5 The H/K PCFI of the fuzzy priorities of criterion i derived by K decision makers, indicated by $\tilde{w}_i(1) \sim \tilde{w}_i(K)$, is denoted by $\widetilde{PCFI}^{H/K}(\{\tilde{w}_i(k)\})$ such that

$$\mu_{\widetilde{PCFI}^{H/K}}(x) = \max_g(\min(\mu_{\tilde{w}_i(g(1))}(x), \ldots, \mu_{\tilde{w}_i(g(H))}(x))) \tag{4.27}$$

where $g() \in Z^+$; $1 \leq g() \leq K$; $g(p) \cap g(q) = \varnothing \; \forall \, p \neq q$; $H \geq 2$.

If $\widetilde{PCFI}^{H/K}(\{\tilde{w}_i(k)\}) = \varnothing$, the partial consensus does not exist.

Example 4.3 The fuzzy priorities of a criterion derived by three decision makers are

Decision maker #1: (0.048, 0.134, 0.246)
Decision maker #2: (0.163, 0.284, 0.526)
Decision maker #3: (0.262, 0.376, 0.612)

The FI of the fuzzy priorities is shown in Fig. 4.7. Obviously, an overall consensus has not been reached by all decision makers. Therefore, the partial consensus among most (i.e., two) decision makers is to be sought. The PCFI result of the fuzzy priorities is shown in Fig. 4.8. As an alternative, the FWI result of the fuzzy priorities is shown in Fig. 4.9, for which the authority levels of decision makers are 0.45, 0.32, and 0.23, respectively. In sum,

(1)   If these decision makers have equal authority levels, the overall consensus among them does not exist.

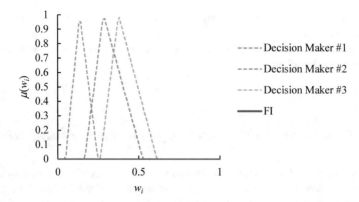

Fig. 4.7 FI result

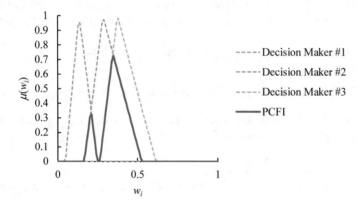

Fig. 4.8 PCFI result

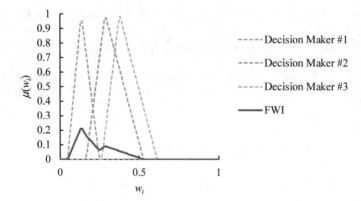

Fig. 4.9 FWI result

(2)   If the authority levels of these decision makers are unequal, the overall consensus, in terms of the FWI result, exists.
(3)   A partial consensus among these decision makers exists, whether their authority levels are equal or not.

4.2.3 Consensus on the Overall Performances of Alternatives

When the overall performances of alternatives are given in numbers, they may or may not overlap. The aforementioned measures can all be applied to check the existence of an overall (or partial) consensus among decision makers [19].

Definition 4.6 The FI of the overall performances of alternative q evaluated by K decision makers, indicated by $\tilde{O}_q(1) \sim \tilde{O}_q(K)$, is denoted by $\widetilde{FI}(\{\tilde{O}_q(k)\})$ such that

$$\mu_{\widetilde{FI}(\{\tilde{O}_q(k)\})}(x) = \min_k(\{\mu_{\tilde{O}_q(k)}(x)\}) \tag{4.28}$$

If $\widetilde{FI}(\{\tilde{O}_q(k)\}) = \emptyset$, the overall consensus does not exist.

Definition 4.7 The FWI of the overall performances of alternative q evaluated by K decision makers, indicated by $\tilde{O}_q(\tilde{O}_q) \sim \tilde{O}_q(K)$, is denoted by $\widetilde{FWI}(\{\tilde{O}_q(k)\})$ such that

$$\mu_{\widetilde{FWI}}(x) = \min_l \mu_{\tilde{O}_q(l)}(x) + \sum_m (\omega_m - \min_l \omega_l)(\mu_{\tilde{O}_q(m)}(x) - \min_l \mu_{\tilde{O}_q(l)}(x)) \tag{4.29}$$

where ω_m is the authority level (or weight) of decision maker m; $\omega_{m_1} \neq \omega_{m_2} \; \exists m_1 \neq m_2$; $\sum_m \omega_m = 1$.
   If $\widetilde{FWI}(\{\tilde{O}_q(k)\}) = \emptyset$, the overall consensus does not exist.

Definition 4.8 The H/K PCFI of the overall performances of alternative q evaluated by K decision makers, indicated by $\tilde{O}_q(1) \sim \tilde{O}_q(K)$, is denoted by $\widetilde{PCFI}^{H/K}(\{\tilde{O}_q(k)\})$ such that

$$\mu_{\widetilde{PCFI}^{H/K}}(x) = \max_g(\min(\mu_{\tilde{O}_q(g(1))}(x), \; ..., \; \mu_{\tilde{O}_q(g(H))}(x))) \tag{4.30}$$

where $g() \in Z^+$; $1 \leq g() \leq K$; $g(p) \cap g(q) = \emptyset \; \forall p \neq q$; $H \geq 2$.
   If $\widetilde{PCFI}^{H/K}(\{\tilde{O}_q(k)\}) = \emptyset$, the partial consensus does not exist.

Example 4.4 Three decision makers evaluated the overall performances of an alternative as

Decision maker #1: (3.67, 4.22, 4.78)
Decision maker #2: (2.89, 3.45, 4.12)
Decision maker #3: (3.11, 3.79, 4.55)

The FI, PCFI, and FWI results of the fuzzy overall performances are shown in
Figs. 4.10, 4.11 and 4.12, respectively. The authority levels of the decision makers are
0.26, 0.33, and 0.41, respectively. Obviously, there is an overall (or partial) consensus
among these decision makers, regardless of the consensus measurement method.

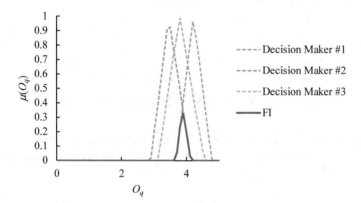

Fig. 4.10 FI result

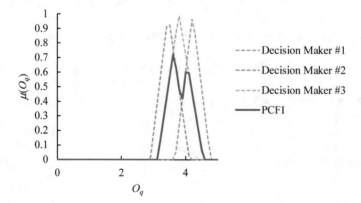

Fig. 4.11 PCFI result

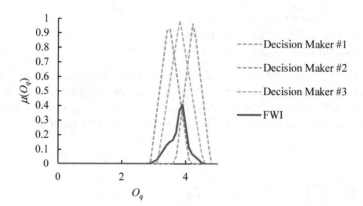

Fig. 4.12 FWI result

4.2.4 Consensus on the Preferences for Alternatives

The preference of decision maker k for alternative q (indicated by T_q) to alternative r (indicated by T_r) can be derived according to the extension principle [20], as illustrated by Fig. 4.13:

$$\mu_{\tilde{S}_k}(T_q > T_r) = \max_{x_q > x_r}(\min(\mu_{\tilde{O}_q}(x_q),\ \mu_{\tilde{O}_r}(x_r))) \tag{4.31}$$

where $\tilde{S}_k = \{(T_q > T_r,\ \mu_{\tilde{S}_k}(T_q > T_r))\}$ is the set of preferences of decision maker k.

Therefore, there is a consensus among decision makers if $FI(\{\tilde{S}_k\}) \neq \emptyset$:

$$FI(\{\tilde{S}_k\}) = \{(T_q > T_r,\ \min_k \mu_{\tilde{S}_k}(T_q > T_r))\} \tag{4.32}$$

Property 4.1 $\mu_{\tilde{S}_k}(T_q > T_r) = 1$ if $O_{q2} \geq O_{r2}$.

Fig. 4.13 Comparing the preferences for alternatives

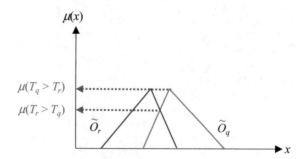

Table 4.2 Overall performances of three alternatives evaluated by three decision makers

Decision maker	$\tilde{O}_1$	$\tilde{O}_2$	$\tilde{O}_3$
#1	(2.49, 3.15, 4.07)	(3.01, 3.67, 4.42)	(2.57, 3.03, 3.58)
#2	(2.67, 3.23, 4.11)	(2.98, 3.55, 4.22)	(1.98, 2.73, 3.39)

Example 4.5 The overall performances of three alternatives evaluated by two decision makers are summarized in Table 4.2. Decision makers' preferences for these alternatives are derived as follows

Decision maker #1:

$$\tilde{S}_1 = \{(T_1 > T_2, \ 0.66), \ (T_2 > T_1, \ 1), \ (T_1 > T_3, \ 1),$$
$$(T_3 > T_1, \ 0.87), \ (T_2 > T_3, \ 1), \ (T_3 > T_2, \ 0.44)\}$$

Decision maker #2:

$$\tilde{S}_2 = \{(T_1 > T_2, \ 0.74), \ (T_2 > T_1, \ 1), \ (T_1 > T_3, \ 1),$$
$$(T_3 > T_1, \ 0.59), \ (T_2 > T_3, \ 1), \ (T_3 > T_2, \ 0.29)\}$$

Therefore,

$$FI(\{\tilde{S}_k\}) = \{(T_1 > T_2, \ 0.66), \ (T_2 > T_1, \ 1), \ (T_1 > T_3, \ 1),$$
$$(T_3 > T_1, \ 0.59), \ (T_2 > T_3, \ 1), \ (T_3 > T_2, \ 0.29)\}$$

$FI(\{\tilde{S}_k\}) \neq \emptyset$. Therefore, there is a consensus between the two decision makers.

4.3   Consensus Enhancement

Kahraman et al. [21] mentioned the fuzzy synthetic evaluation method in which the performance of an alternative in optimizing a criterion is evaluated by each decision maker in several linguistic terms. For example, the performance of alternative q in criterion i evaluated by decision maker k is

"Excellent" with a membership of 0.7
"Good" with a membership of 0.55
"Poor" with a membership of 0.35

Since the same linguistic terms are used by all decision makers, the evaluation results by them will overlap, ensuring the existence of an overall consensus.

The decision maker whose fuzzy judgment matrix, fuzzy priorities of criteria, or overall performances of alternatives is farthest from the average is either excluded from the fuzzy group decision-making process [22] or asked to modify his/her fuzzy judgment matrix, fuzzy priorities of criteria, or overall performances of alternatives [23]. Modifying a fuzzy judgment matrix is less controversial, while modifying the others may be questionable. An example is given below.

Example 4.5 The fuzzy judgment matrixes of three decision makers are as follows:

$$
\tilde{A}(1) = \begin{vmatrix} 1 & (1,\ 2,\ 4) & - \\ - & 1 & - \\ (1,\ 3,\ 5) & (3,\ 5,\ 7) & 1 \end{vmatrix}
$$

$$
\tilde{A}(2) = \begin{vmatrix} 1 & (3,\ 5,\ 7) & - \\ - & 1 & - \\ (2,\ 4,\ 6) & (1,\ 2,\ 4) & 1 \end{vmatrix}
$$

$$
\tilde{A}(3) = \begin{vmatrix} 1 & (1,\ 3,\ 5) & - \\ - & 1 & - \\ (1,\ 2,\ 4) & (4,\ 6,\ 8) & 1 \end{vmatrix}
$$

After aggregating, $\widetilde{FI}(\{\tilde{A}(k)\}) = \emptyset$, showing a lack of consensus among decision makers. To solve this problem, the average of fuzzy judgment matrixes is calculated as

$$
\overline{\tilde{A}} = \begin{vmatrix} 1 & (1.67,\ 3.33,\ 5.33) & - \\ - & 1 & - \\ (1.33,\ 3,\ 5) & (2.67,\ 4.33,\ 6.33) & 1 \end{vmatrix}
$$

The distance from each fuzzy judgment matrix to the average is measured according to Eq. (4.7) as

$$
dist\,(\tilde{A}(1),\ \overline{\tilde{A}}) = 1.49
$$

$$
dist\,(\tilde{A}(2),\ \overline{\tilde{A}}) = 3.04
$$

$$
dist\,(\tilde{A}(3),\ \overline{\tilde{A}}) = 1.97
$$

Obviously, the fuzzy judgment matrix of Decision Maker #2 is farthest from the average. Decision Maker #2 is asked to modify his fuzzy judgment matrix to be closer to the average:

$$\text{Modified } \tilde{A}(2) = \begin{vmatrix} 1 & (3, 5, 7) & - \\ - & 1 & - \\ (2, 4, 6) & (2, 3, 5) & 1 \end{vmatrix}$$

Now $\widetilde{FI}(\{\tilde{A}(k)\}) \neq \emptyset$, showing an overall consensus among these decision makers.

However, the fuzzy priorities of criteria are systematically derived from a fuzzy judgment matrix, and cannot be arbitrarily modified. The overall performance of an alternative is evaluated according to the derived fuzzy priorities of criteria, modifying which will face the same problem.

Capuano et al. [23] established a mechanism in which a decision maker adjusts his/her preferences for alternatives by considering the preferences of others that are considered authoritative by him/her, which is beneficial to reaching a partial consensus because less authoritative decision makers will eventually be excluded. Unlike in the FWI operator, the authority levels of decision makers in the mechanism are dynamic.

References

1. F. Herrera, E. Herrera-Viedma, A model of consensus in group decision making under linguistic assessments. Fuzzy Sets Syst. **78**(1), 73–87 (1996)
2. E. Herrera-Viedma, G. Pasi, A.G., Lopez-Herrera, C. Porcel, Evaluating the information quality of web sites: a methodology based on fuzzy computing with words. J. Am. Soc. Inf. Sci. Technol. **57**(4), 538–549 (2006)
3. F.J. Cabrerizo, J.M. Moreno, I.J. Pérez, E. Herrera-Viedma, Analyzing consensus approaches in fuzzy group decision making: advantages and drawbacks. Soft. Comput. **14**(5), 451–463 (2010)
4. T. Chen, Y.C. Wang, A calibrated piecewise-linear FGM approach for travel destination recommendation during the COVID-19 pandemic. Appl. Soft Comput. 107535 (2021)
5. Y.C. Lin, T. Chen, A multibelief analytic hierarchy process and nonlinear programming approach for diversifying product designs: smart backpack design as an example. Proc. Inst. Mech. Eng. Part B: J. Eng. Manuf. **234**(6–7), 1044–1056 (2020)
6. G. Zheng, N. Zhu, Z. Tian, Y. Chen, B. Sun, Application of a trapezoidal fuzzy AHP method for work safety evaluation and early warning rating of hot and humid environments. Saf. Sci. **50**(2), 228–239 (2012)
7. T. Chen, A diversified AHP-tree approach for multiple-criteria supplier selection. Comput. Manag. Sci. 1–23 (2021)
8. A.I. Ölçer, A.Y. Odabaşi, A new fuzzy multiple attributive group decision making methodology and its application to propulsion/manoeuvring system selection problem. Eur. J. Oper. Res. **166**(1), 93–114 (2005)
9. Z. Zhang, X. Chu, Fuzzy group decision-making for multi-format and multi-granularity linguistic judgments in quality function deployment. Expert Syst. Appl. **36**(5), 9150–9158 (2009)
10. L. Yu, S. Wang, K.K. Lai, An intelligent-agent-based fuzzy group decision making model for financial multicriteria decision support: the case of credit scoring. Eur. J. Oper. Res. **195**(3), 942–959 (2009)

11. T. Chen, Y.C. Lin, A fuzzy-neural system incorporating unequally important expert opinions for semiconductor yield forecasting. Int. J. Uncertain. Fuzziness Knowl. Based Syst. **16**(01), 35–58 (2008)
12. T., Chen, H. C. Wu, Fuzzy collaborative intelligence fuzzy analytic hierarchy process approach for selecting suitable three-dimensional printers. Soft Comput. **25**(5), 4121–4134 (2021)
13. T. Chen, Y.-C. Wang, M.-C. Chiu, Assessing the robustness of a factory amid the COVID-19 pandemic: a fuzzy collaborative intelligence approach. Healthcare **8**, 481 (2020)
14. T. Chen, Y.-C. Wang, C.-W. Lin, A fuzzy collaborative forecasting approach considering experts' unequal levels of authority. Appl. Soft Comput. **94**, 106455 (2020)
15. H.-C. Wu, T.-C.T. Chen, C.-H. Huang, Y.-C. Shi, Comparing built-in power banks for a smart backpack design using an auto-weighting fuzzy-weighted-intersection FAHP approach. Mathematics **8**(10), 1759 (2020)
16. T. Chen, A hybrid fuzzy and neural approach with virtual experts and partial consensus for DRAM price forecasting. Int. J. Innov. Comput. Inf. Control. **8**, 583–597 (2012)
17. M.-C. Chiu, T. Chen, Assessing mobile and smart technology applications to active and healthy ageing using a fuzzy collaborative intelligence approach. Cognit. Comput. **13**(2), 431–446 (2021)
18. H.-C. Wu, Y.-C. Wang, T. Chen, Assessing and comparing COVID-19 intervention strategies using a varying partial-consensus fuzzy collaborative intelligence approach. Mathematics **8**, 1725 (2020)
19. T. Chen, Y.-C. Wang, H.-C. Wu, Analyzing the impact of vaccine availability on alternative supplier selection amid the COVID-19 pandemic: a cFGM-FTOPSIS-FWI approach. Healthcare **9**(1), 71 (2021)
20. R.R. Yager, A characterization of the extension principle. Fuzzy Sets Syst. **18**(3), 205–217 (1986)
21. C. Kahraman, D. Ruan, I. Doğan, Fuzzy group decision-making for facility location selection. Inf. Sci. **157**, 135–153 (2003)
22. H. Gao, Y. Ju, E.D.S. Gonzalez, W. Zhang, Green supplier selection in electronics manufacturing: an approach based on consensus decision making. J. Clean. Prod. 118781 (2019)
23. N. Capuano, F. Chiclana, H. Fujita, E. Herrera-Viedma, V. Loia, Fuzzy group decision making with incomplete information guided by social influence. IEEE Trans. Fuzzy Syst. **26**(3), 1704–1718 (2017)

Chapter 5
Aggregation Mechanisms

5.1 Aggregation Time Points

Only after decision makers reach an overall (or partial) consensus, their judgments or decisions can be aggregated. However, the aggregation mechanism may be different from the consensus measurement method.

Using fuzzy sets to express decision makers' judgments (or decisions) facilitates the subsequent aggregation, because these judgments (or decisions) may overlap, meaning that a judgment (or decision) acceptable to all decision makers can be found [1]. Aggregating decision makers' judgments (or decisions) is usually done by averaging these judgments (or decisions) or finding their overlapping parts. A suitable aggregation mechanism is easy and efficient to implement and communicate, and can generate a reasonable aggregation result [2–5].

There are several time points in a fuzzy group decision-making process by which the judgments (or decisions) of decision makers can be aggregated, as illustrated in Fig. 5.1:

(1) Before comparing the relative priorities of criteria;
(2) After comparisons are made, but before the priorities of criteria are derived;
(3) After deriving the priorities of criteria, but before evaluating the overall performances of alternatives;
(4) After evaluating the (overall) performances of alternatives, but before ranking alternatives (or expressing preferences for alternatives);
(5) After expressing preferences for alternatives.

In a fuzzy compromise multiple criteria decision-making (MCDM) method [6], the judgements (or pairwise comparison results) of decision makers, the derived priorities of criteria, and the evaluated overall performances of alternatives can be aggregated. In a fuzzy outranking MCDM method [7], the priorities of criteria and the preferences for alternatives can be aggregated.

T.-C. T. Chen, *Advances in Fuzzy Group Decision Making*,
SpringerBriefs in Applied Sciences and Technology,
https://doi.org/10.1007/978-3-030-86208-4_5

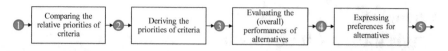

Fig. 5.1 Various aggregation time points in a fuzzy group decision-making process

Some references are reviewed as follows. Herrera and Herrera-Viedma [8] proposed the linguistic ordered weighted average (LOWA) operator to aggregate decision makers' preferences for various alternatives. The LOWA operator can also be applied to aggregate decision makers' judgments on the relative priorities of criteria. In Kahraman et al. [9], several fuzzy group decision-making methods were applied to a facility location selection problem. If fuzzy relations were applied, decision makers' decisions (i.e., their preferences for various alternatives) were aggregated by calculating the percentage of the preference for each alternative. If the fuzzy analytic hierarchy process (FAHP) was applied, decision makers' judgments on the relative priorities of criteria were aggregated using fuzzy geometric mean (FGM). Boran et al. [10] also dealt with a facility location selection problem, in which the performances of alternatives evaluated by decision makers were aggregated using fuzzy weighted average (FWA). Wang and Elhag [11] also applied FWA to aggregate decision makers' judgments on the relative priorities of criteria and the overall performances of alternatives evaluated by them. Capuano et al. [12] applied the ordered weighted average (OWA) operator to aggregate decision makers' preferences for various alternatives. To find out the critical factors for applying advanced three-dimensional (3D) printing technologies to the aircraft industry, Wang et al. [13] applied an FAHP method, in which FGM was applied to aggregate decision makers' judgments on the relative priorities of criteria. Pishdar et al. [14] proposed a fuzzy best–worst method (fuzzy BWM) to choose a suitable hub airport. In the fuzzy BWM method, each decision maker first chose the most important and the least important criteria. Then, the other criteria were compared with the two criteria in pairs. In this way, the number of pairwise comparisons could be reduced. Finally, a mathematical programming problem was solved to derive the priorities of criteria. The priorities of criteria derived by all decision makers were averaged. Chen et al. [15] proposed a fuzzy collaborative intelligence method to assessing the robustness of a factory amid the COVID-19 pandemic. If there was overall consensus among decision makers, fuzzy intersection (FI) was applied to aggregate the priorities of criteria derived by them; otherwise, partial-consensus FI (PCFI), which was the consensus among most decision makers, was applied instead. In recent years, a number of fuzzy OWA operators have been devised to aggregate the performances of an alternative in different scenarios or in optimizing various criteria, e.g., fuzzy generalized OWA operator [16], intuitionistic FGOWA operator [17], induced generalized intuitionistic fuzzy OWA operator [18], Pythagorean fuzzy probabilistic OWA operator [19], fuzzy-stochastic OWA operator [20], and dynamic fuzzy OWA operator [21], which provides a sophisticated alternative to the commonly applied FWA method. However, the focus is on the aggregation of criteria rather than the aggregation of decision makers. In addition, the priorities of criteria are determined in an objective and systematic way. Although this treatment reduces the load on decision makers, it also hampers their subjectivity. In addition, some of the more advanced fuzzy OWA operators are not easy to understand and communicate.

In theory, the earlier the decision makers' judgments (or decisions) are aggregated, the easier the subsequent operations will become. However, premature aggregation may hinder the subjectivity of decision makers. Therefore, there is a need to make trade-offs. In most of the previous studies, decision makers' judgments were aggregated before deriving the priorities of criteria.

5.2 Aggregating Decision Makers' Judgments

5.2.1 Fuzzy (Weighted) Geometric Mean, Fuzzy Arithmetic Mean, and Fuzzy Weighted Average

FGM has been widely applied to aggregate decision makers' judgments in FAHP, which is based on a ratio scale. In contrast, the application of fuzzy arithmetic mean (FAM) to FAHP may lead to unreasonable results [13]. FAM is a suitable aggregator for methods based on an interval scale, such as fuzzy measuring attractiveness by a categorically based evaluation technique (fuzzy MACBETH) [22].

Assume that the pairwise comparison results by decision maker k are placed in a fuzzy judgment matrix $\tilde{\mathbf{A}}(k) = [\tilde{a}_{ij}(k)]$ where

$$\tilde{a}_{ji}(k) = \begin{cases} 1 & if \quad j = i \\ 1/\tilde{a}_{ij}(k) & otherwise \end{cases} ; i, j \in [1, n] \tag{5.1}$$

$k = 1 \sim K$. The aggregation result using FGM is

$$\tilde{a}_{ij} = \sqrt[K]{\prod_{k=1}^{K} \tilde{a}_{ij}(k)}; i, j = 1 \sim n \tag{5.2}$$

If decision makers have unequal authority levels (or weights), fuzzy weighted geometric mean (FWGM) is applicable:

$$\tilde{a}_{ij} = \prod_{k=1}^{K} \tilde{a}_{ij}(k)^{\omega_k} \tag{5.3}$$

where ω_k is the authority level of decision maker k. $\sum_{k=1}^{K} \omega_k = 1$.

The aggregation result using FAM is [23]

$$\tilde{a}_{ij} = \frac{\sum_{k=1}^{K} \tilde{a}_{ij}(k)}{K} \tag{5.4}$$

When decision makers have unequal authority levels (or weights), FWA can be applied:

$$\tilde{a}_{ij} = \frac{\sum_{k=1}^{K} (\omega_k \tilde{a}_{ij}(k))}{\sum_{k=1}^{K} \omega_k}$$

$$= \sum_{k=1}^{K} (\omega_k \tilde{a}_{ij}(k)) \tag{5.5}$$

since $\sum_{k=1}^{K} \omega_k = 1$. FWA has also been applied to aggregate decision makers' preferences for various alternatives when decision makers have unequal authority levels in methods based on fuzzy relationships [24].

5.2.2 Linguistic Ordered Weighted Average

Unlike the original judgments, the aggregation result is no longer an integer within [1, 9], which may be a concern of decision makers. The LOWA operator [8] can be applied to solve this problem, which is composed of the following steps:

(1) Index linguistic (or semantic) terms: $S = \{s_0, \ldots, s_G\}$; $s_g \leq s_{g+1}$.
(2) Sort the choices (i.e., judgments) of decision makers descendingly: Consider the case of two decision makers k and l who chose s_k and s_l, respectively. Assuming $s_k \leq s_l$, then the sorting result is $\{s_l, s_k\}$.
(3) Assign descending weights to the sorted choices: $\{w_1, w_2\}$; $w_1 \geq w_2$; $w_1 + w_2 = 1$.
(4) Derive the LOWA result s_p as

$$s_p = C^2\{w_1, w_2, s_l, s_k\} \tag{5.6}$$

where

$$p = \min(G, k + \text{roundup}(w_1(l - k)) \tag{5.7}$$

Now consider the case of three decision makers k, l, and m. Assume $s_m \leq s_k \leq s_l$. The sorting result is $\{s_l, s_k, s_m\}$. The LOWA result can be derived as

$$s_p = C^3\{w_1, w_2, w_3, s_l, s_k, s_m\}$$

$$= C^2\{w_1, (1 - w_1), s_l, C^2\{\frac{w_2}{w_2 + w_3}, \frac{w_3}{w_2 + w_3}, s_k, s_m\}\} \tag{5.8}$$

The case of more than three decision makers can be handled similarly. An example is given below.

Example 5.1 In an FAHP problem, the linguistic terms are indexed from s_0 (as equal as) to s_8 (absolutely more important than). Therefore, $G = 8$. The fuzzy judgment matrixes of three decision makers are as follows:

$$\tilde{A}(1) = \begin{vmatrix} 1 & s_2 & 1/s_3 \\ 1/s_2 & 1 & 1/s_5 \\ s_3 & s_5 & 1 \end{vmatrix}$$

$$\tilde{A}(2) = \begin{vmatrix} 1 & s_5 & 1/s_4 \\ 1/s_5 & 1 & 1/s_2 \\ s_4 & s_2 & 1 \end{vmatrix}$$

$$\tilde{A}(3) = \begin{vmatrix} 1 & s_3 & 1/s_3 \\ 1/s_3 & 1 & 1/s_6 \\ s_3 & s_6 & 1 \end{vmatrix}$$

For example, to derive the value of $\tilde{a}_{12}$, the choices of decision makers are sorted as $\{s_5, s_3, s_2\}$. Assume the moderately optimistic decision strategy is applied, which assigns descending weights 0.72, 0.17, and 0.11. Then, the LOWA result is derived as

$$s_p = C^3\{0.72, 0.17, 0.11, s_5, s_3, s_2\}$$
$$= C^2\{0.72, 0.28, s_5, C^2\{\frac{0.17}{0.17 + 0.11}, \frac{0.11}{0.17 + 0.11}, s_3, s_2\}\}$$

The following subproblem is solved first:

$$s_p = C^2\{\frac{0.17}{0.17 + 0.11}, \frac{0.11}{0.17 + 0.11}, s_3, s_2\}$$

According to Eq. (5.7),

$$p = \min(8, \ 2 + \text{roundup}(0.61(3 - 2)) = 3$$

Therefore, the problem becomes

$$s_p = C^2\{0.72, 0.28, s_5, s_3\}$$

where

$$p = \min(8, \ 3 + \text{roundup}(0.72(5 - 3)) = 5$$

Therefore, $\tilde{a}_{12} = s_5$. Finally, the aggregation result is

$$\tilde{A} = \begin{vmatrix} 1 & s_5 & 1/s_4 \\ 1/s_5 & 1 & 1/s_6 \\ s_4 & s_6 & 1 \end{vmatrix}$$

A similar yet simpler method was proposed by Lyu et al. [25]. First, each criterion is given a score. Assume the score given to criterion i by decision maker k is indicated by $\xi_i(k)$; $\xi_i(k) \in [1, 9]$. Then, $\tilde{a}_{ij} = s_p$ (i.e., the pth linguistic term) where

$$p = \text{roundup}(\text{mod}(\{\xi_i(k)\})/ \text{ mod } (\{\xi_j(k)\})) \tag{5.9}$$

If the result of mod() is not unique,

$$p = \text{roundup}(\text{avg}(\{\xi_i(k)\})/\text{avg}(\{\xi_j(k)\})) \tag{5.10}$$

5.3   Aggregating the Priorities of Criteria Derived by Decision Makers

5.3.1   Fuzzy Arithmetic Mean, Fuzzy Weighted Average, and Fuzzy (Weighted) Geometric Mean

FAM is a prevalent method to aggregate the priorities of criteria derived by decision makers:

$$\tilde{w}_i = \frac{\sum_{k=1}^{K} \tilde{w}_i(k)}{K}; i = 1 \sim n \tag{5.11}$$

FAM is easy to calculate and does not need to normalize the aggregation result. If decision makers have unequal authority levels (or weights), FWA is applicable:

$$\tilde{w}_i = \sum_{k=1}^{K} (\omega_k \tilde{w}_i(k)) \tag{5.12}$$

However, the aggregation result needs to be normalized:

$$\tilde{w}_i \rightarrow \frac{\tilde{w}_i}{\sum_{j=1}^{n} \tilde{w}_j} \tag{5.13}$$

Theorem 5.1 *Equation (5.11)can be rewritten as*

$$\tilde{w}_i = \left(\frac{1}{1 + \sum_{j \neq i} \frac{w_{j3}}{w_{i1}}}, \frac{1}{1 + \sum_{j \neq i} \frac{w_{j2}}{w_{i2}}}, \frac{1}{1 + \sum_{j \neq i} \frac{w_{j1}}{w_{i3}}} \right) \quad (5.14)$$

FGM is also applicable:

$$\tilde{w}_i = \sqrt[K]{\prod_{k=1}^{K} \tilde{w}_i(k)}; i = 1 \sim n \quad (5.15)$$

If decision makers have unequal authority levels (or weights), FWGM can be applied [26]:

$$\tilde{w}_i = \prod_{k=1}^{K} \tilde{w}_i(k)^{\omega_k} \quad (5.16)$$

The aggregation result also needs to be normalized.

5.3.2   Fuzzy Intersection, Fuzzy Weighted Intersection, and Partial-Consensus Fuzzy Intersection

FI [27] has also been applied to aggregate the priorities of criteria derived by decision makers.

Definition 5.1 The FI of the priorities of criteria derived by K decision makers for the ith criterion, indicated by $\tilde{w}_i(1) \sim \tilde{w}_i(K)$, is denoted by $\widetilde{FI}(\tilde{w}_i(1), \ldots, \tilde{w}_i(K))$ such that

$$\mu_{\widetilde{FI}}(x) = \min_k \{\mu_{\tilde{w}_i(k)}(x)\} \quad (5.17)$$

An example is provided in Fig. 5.2.

When the priorities of criteria derived by decision makers are given in triangular fuzzy numbers (TFNs), the FI result is a polygonal fuzzy number that is not easy to be used in subsequent operations. For solving this problem, the FI result can be approximated with an equivalent TFN as [28]

$$\widetilde{FI}(\{\tilde{w}_i(k)\}) \cong (\min(\widetilde{FI}(\{\tilde{w}_i(k)\}))$$
$$3COG(\widetilde{FI}(\{\tilde{w}_i(k)\})) - \min(\widetilde{FI}(\{\tilde{w}_i(k)\})) - \max(\widetilde{FI}(\{\tilde{w}_i(k)\}))$$
$$\max(\widetilde{FI}(\{\tilde{w}_i(k)\}))) \quad (5.18)$$

where $COG()$ is the center-of-gravity function [29]:

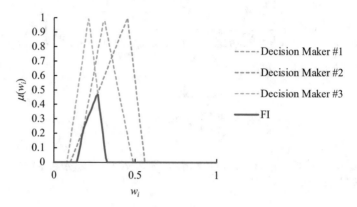

Fig. 5.2 An example of aggregating the priorities of criteria derived by decision makers using FI

$$COG(\widetilde{FI}(\{\tilde{w}_i(k)\})) = \frac{\int_{all\ x} x\mu_{\widetilde{FI}}(x)dx}{\int_{all\ x} \mu_{\widetilde{FI}}(x)dx} \tag{5.19}$$

as illustrated in Fig. 5.3. In this way, the minimum, COG (i.e., the defuzzification result), and maximum of the equivalent TFN are equal to those of the FI result.

Example 5.2 The priorities of a criterion derived by three decision makers are

Decision maker #1: (0.141, 0.287, 0.444).
Decision maker #2: (0.105, 0.434, 0.629).
Decision maker #3: (0.067, 0.162, 0.370).

The FI result of the derived priorities is shown in Fig. 5.4. For example, the membership of 0.2 in the FI result is min(0.40, 0.29, 0.82) = 0.29. The FI result is a polygonal fuzzy number with a minimum of 0.141 and a maximum of 0.370. The COG of the FI result is calculated using Eq. (5.19) as 0.254. Therefore, the FI

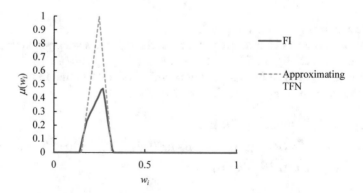

Fig. 5.3 Approximating the FI result with a TFN

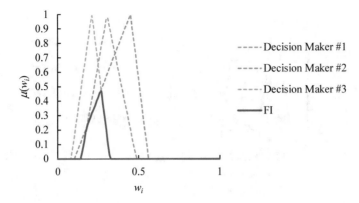

Fig. 5.4 The FI result

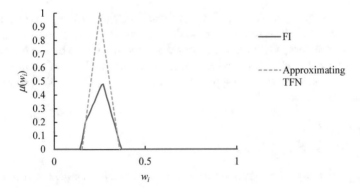

Fig. 5.5 The equivalent TFN

result can be approximated with a TFN (0.141, 3 · 0.254 – 0.141 – 0.370, 0.370), as illustrated in Fig. 5.5.

If decision makers have unequal authority levels (or weights), the fuzzy weighted intersection (FWI) operator proposed by Chen et al. [30] is applicable.

Definition 5.2 The FWI of the priorities of criterion i derived by K decision makers, indicated by $\tilde{w}_i(1) \sim \tilde{w}_i(K)$, is denoted by $\widetilde{FWI}(\tilde{w}_i(1), \ldots, \tilde{w}_i(K))$ such that

$$\mu_{\widetilde{FWI}}(x) = \min_l \mu_{\tilde{w}_i(l)}(x) + \sum_k (\omega_k - \min_l \omega_l)(\mu_{\tilde{w}_i(k)}(x) - \min_l \mu_{\tilde{w}_i(l)}(x)) \quad (5.20)$$

where ω_k is the authority level (or weight) of decision maker k; $\omega_k \neq \omega_l \; \exists k \neq l$; $\sum_k \omega_k = 1$.

If $\omega_k = {}^1/_K$ for all k, FWI reduces to FI. An example is provided in Fig. 5.6, which illustrates the aggregation of the priorities of criteria derived by three decision

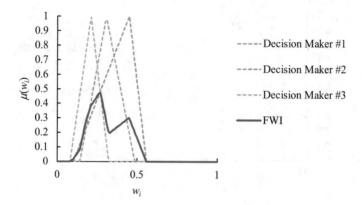

Fig. 5.6 An example of aggregating the priorities of criteria using FWI

makers. In this figure, the authority level of Decision maker #2 is higher than those of the others, and the FWI result is closer to the priority derived by Decision maker #2. Specifically speaking, values that are considered more possible by Decision maker #2 or all decision makers will have higher memberships in the FWI result.

The FWI result can also be approximated with an equivalent TFN to facilitate subsequent operations [31]:

$$
\widetilde{FWI}(\{\tilde{w}_i(k)\})
$$
$$
\cong (\min(\widetilde{FWI}(\{\tilde{w}_i(k)\})))
$$
$$
3COG(\widetilde{FWI}(\{\tilde{w}_i(k)\})) - \min(\widetilde{FWI}(\{\tilde{w}_i(k)\})) - \max(\widetilde{FWI}(\{\tilde{w}_i(k)\}))
$$
$$
\max(\widetilde{FWI}(\{\tilde{w}_i(k)\}))) \tag{5.21}
$$

as illustrated in Fig. 5.7.

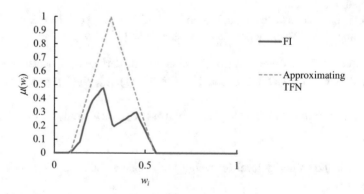

Fig. 5.7 Approximating the FWI result with a TFN

In Example 5.2, assume the authority levels of the three decision makers are 0.45, 0.33, and 0.22, respectively. The FWI result of the derived priorities is shown in Fig. 5.8. For example, the membership of 0.2 in the FWI result is min(0.40, 0.29, 0.82) + (0.45 − 0.22) · (0.40 − 0.29) + (0.33 − 0.22) · (0.29 − 0.29) + (0.22 − 0.22) · (0.82 − 0.29) = 0.32. The FWI result has a minimum of 0.105 and a maximum of 0.629. The COG of the FWI result is 0.303. Therefore, the FWI result can be approximated with a TFN (0.105, 0.180, 0.629), as illustrated in Fig. 5.9.

When there is only a partial consensus among most decision makers, the PCFI operator [28, 32, 33] is applied to aggregate the derived priorities instead.

Definition 5.3 The H/K partial-consensus fuzzy intersection (PCFI) of the priorities of criterion i derived by K decision makers, indicated by $\tilde{w}_i(1) \sim \tilde{w}_i(K)$, is denoted by $\widetilde{PCFI}^{H/K}(\{\tilde{w}_i(k)\})$ such that

$$\mu_{\widetilde{PCFI}^{H/K}}(x) = \max_g(\min(\mu_{\tilde{w}_i(g(1))}(x), \ldots, \mu_{\tilde{w}_i(g(H))}(x))) \qquad (5.22)$$

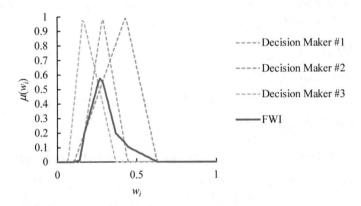

Fig. 5.8 The FWI result

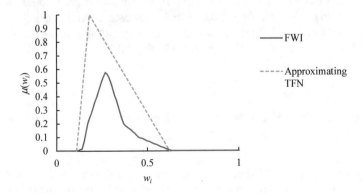

Fig. 5.9 The approximating TFN

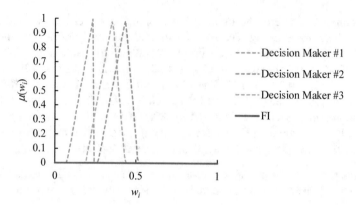

Fig. 5.10 The FI result is an empty set

where $g() \in Z^+; 1 \leq g() \leq K; g(p) \cap g(q) = \emptyset \; \forall \, p \neq q; H \geq 2$.

The PCFI result can be approximated with an equivalent TFN to facilitate subsequent operations:

$$
\begin{aligned}
&\widehat{PCFI}(\{\tilde{w}_i(k)\}) \\
&\cong (\min(\widehat{PCFI}(\{\tilde{w}_i(k)\}))) \\
&3COG(\widehat{PCFI}(\{\tilde{w}_i(k)\})) - \min(\widehat{PCFI}(\{\tilde{w}_i(k)\})) - \max(\widehat{PCFI}(\{\tilde{w}_i(k)\})) \\
&\max(\widehat{PCFI}(\{\tilde{w}_i(k)\})))
\end{aligned}
\tag{5.23}
$$

Example 5.3 Three decision makers derived the priorities of a criterion as

Decision maker #1: (0.074, 0.230, 0.239).
Decision maker #2: (0.265, 0.433, 0.513).
Decision maker #3: (0.189, 0.354, 0.434).

The FI result is an empty set, showing that the overall consensus does not exist, as shown in Fig. 5.10. Therefore, the partial consensus is sought using the PCFI operator, as illustrated in Fig. 5.11. To facilitate the subsequent operation, the PCFI result is approximated with a TFN as shown in Fig. 5.12.

5.4 Aggregating the (Overall) Performances of Alternatives Evaluated by Decision Makers

In fuzzy compromise MCDM methods, most of the previous studies focused on the aggregation of the pairwise comparison results of decision makers or the priorities of criteria derived by them. Aggregating the overall performances of alternatives

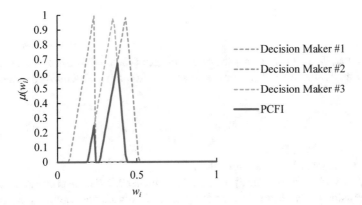

Fig. 5.11   The PCFI result

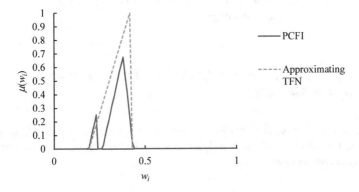

Fig. 5.12   The equivalent TFN for the PCFI result

evaluated by decision makers was less investigated. In contrast, fuzzy outranking MCDM methods need to aggregate decision makers' preferences for alternatives, which is obviously based on the overall performances of alternatives.

In some studies, the performances of an alternative in optimizing a criterion evaluated by decision makers are aggregated. There are also studies in which the overall performances of an alternative evaluated by decision makers are aggregated.

5.4.1   Aggregating the Performances in Optimizing a Criterion Evaluated by Decision Makers

In Zheng et al. [34], decision maker k evaluated the performance of alternative q in optimizing criterion i as $\tilde{f}_{qi}(k)$. Then, the evaluation results by all decision makers were averaged:

$$\tilde{f}_{qi} = \frac{\sum_{k=1}^{K} \tilde{f}_{qi}(k)}{K} \qquad (5.24)$$

However, the result was only the performance of alternative q in optimizing criterion i. Finally, the overall performance of the alternative was derived using FWA as

$$\tilde{O}_q = \sum_{i=1}^{n} (\tilde{w}_i(\times)\tilde{f}_{qi}) \qquad (5.25)$$

where $(\times)$ denotes fuzzy multiplication.

Similarly, in Junior et al. [35], each decision maker derived the priorities of criteria and evaluated the performances of alternatives in optimizing various criteria individually. Then, these results were averaged according to Eq. (5.24) and then fed into the fuzzy technique for order preference by similarity to the ideal solution (fuzzy TOPSIS) method to evaluate the overall performance of the alternative, as if a single decision maker was making the decision.

5.4.2 Aggregating the Overall Performances Evaluated by Decision Makers

Assume that the overall performance of alternative q evaluated by decision maker k is indicated by $\tilde{O}_q(k)$. FAM is a straightforward method to aggregate the evaluation results by all decision makers:

$$\tilde{O}_q = \frac{\sum_{k=1}^{K} \tilde{O}_q(k)}{K}; q = 1 \sim Q \qquad (5.26)$$

If decision makers have unequal authority levels (or weights), FWA is applicable:

$$\tilde{O}_q = \sum_{k=1}^{K} (\omega_k \tilde{O}_q(k)) \qquad (5.27)$$

In Chen et al. [36], each decision maker evaluated and compared the overall performances of alternatives using fuzzy TOPSIS. Then, FWI was applied to aggregate the evaluation results by all decision makers as

$$\tilde{O}_q = \widetilde{FWI}(\{\tilde{O}_q(k)\}) \qquad (5.28)$$

where

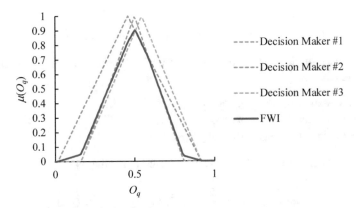

Fig. 5.13 FWI of the overall performances evaluated by three decision makers

$$\mu_{\widetilde{FWI}}(x) = \min_l \mu_{\tilde{O}_q(l)}(x) + \sum_k (\omega_k - \min_l \omega_l)(\mu_{\tilde{O}_q(k)}(x) - \min_l \mu_{\tilde{O}_q(l)}(x)) \quad (5.29)$$

ω_k is the authority level (or weight) of decision maker k. The aggregation result can be defuzzified using COG.

An example is given below.

Example 5.4 The overall performances of an alternative evaluated by three decision makers using fuzzy TOPSIS are

Decision maker #1: $\tilde{O}_q(1) = (0.162,\ 0.502,\ 0.806)$.
Decision maker #2: $\tilde{O}_q(2) = (0.015,\ 0.459,\ 0.915)$.
Decision maker #3: $\tilde{O}_q(3) = (0.160,\ 0.546,\ 0.914)$.

The overall performance of an alternative evaluated using fuzzy TOPSIS is in terms of the fuzzy closeness of the alternative, which is within $[0, 1]$. The authority levels of the three decision makers are 0.35, 0.40, and 0.25, respectively. The FWI of their evaluation results is shown in Fig. 5.13. The defuzzification result using COG is 0.488.

5.4.3 Aggregating Decision Makers' Preferences for Alternatives

Blin [37] proposed an aggregation method based on fuzzy relationships. First, each decision maker ranks alternatives individually. Then, the percentage that alternative q is preferred to alternative r is calculated. All pairwise comparison results are used to develop a fuzzy relationship $\tilde{S}$. The α cut of $\tilde{S}$ includes ordered pairs with memberships higher than α. By specifying a threshold for α, the α' cut of $\tilde{S}$ is used to generate a ranking of alternatives. An example is given below.

Example 5.5 Three decision makers first rank four alternatives, indicated by A_1 to A_4. The results are

Decision maker #1: $A_1 \geq A_3 \geq A_2 \geq A_4$.
Decision maker #2: $A_1 \geq A_2 \geq A_3 \geq A_4$.
Decision maker #3: $A_2 \geq A_1 \geq A_4 \geq A_3$.

The percentage that an alternative is preferred to another is calculated:

$$\mu_{\tilde{S}}(A_1 \geq A_2) = \frac{2}{3} = 0.67; \mu_{\tilde{S}}(A_2 \geq A_1) = 0.33$$
$$\mu_{\tilde{S}}(A_1 \geq A_3) = 1.00$$
$$\mu_{\tilde{S}}(A_1 \geq A_4) = 1.00$$
$$\mu_{\tilde{S}}(A_2 \geq A_3) = 0.67; \mu_{\tilde{S}}(A_3 \geq A_2) = 0.33$$
$$\mu_{\tilde{S}}(A_2 \geq A_4) = 1.00$$
$$\mu_{\tilde{S}}(A_3 \geq A_4) = 0.67; \mu_{\tilde{S}}(A_4 \geq A_3) = 0.33$$

The threshold for α is set to 0.67. As a result, only the following pairwise comparison results hold:

$$A_1 \geq A_2, A_1 \geq A_3, A_1 \geq A_4, A_2 \geq A_3, A_2 \geq A_4, A_3 \geq A_4$$

Only the following ranking result is possible:

$$A_1 \geq A_2 \geq A_3 \geq A_4$$

It is noteworthy that if the threshold for α is set to a higher value, there will be more possible ranking results.

References

1. A. Mardani, M. Nilashi, E.K. Zavadskas, S.R. Awang, H. Zare, N.M. Jamal, Decision making methods based on fuzzy aggregation operators: three decades review from 1986 to 2017. Int. J. Inf. Technol. Decis. Mak. **17**(02), 391–466 (2018)
2. K.P. Lin, K.C. Hung, An efficient fuzzy weighted average algorithm for the military UAV selecting under group decision-making. Knowl.-Based Syst. **24**(6), 877–889 (2011)
3. T. Chen, Enhancing the efficiency and accuracy of existing FAHP decision-making methods. EURO J. Decis. Process. **8**, 177–204 (2020)
4. H.-C. Wu, T. Chen, C.-H. Huang, A piecewise linear FGM approach for efficient and accurate FAHP analysis: smart backpack design as an example. Mathematics **8**, 1319 (2020)

5. T. Chen, Y.-C. Lin, M.-C. Chiu, Approximating alpha-cut operations approach for effective and efficient fuzzy analytic hierarchy process analysis. Appl. Soft Comput. **85**, 105855 (2019)
6. S. Opricovic, A fuzzy compromise solution for multicriteria problems. Int. J. Uncertain. Fuzz. Knowl.-Based Syst. **15**(03), 363–380 (2007)
7. M. Abedi, G.H. Norouzi, N. Fathianpour, Fuzzy outranking approach: a knowledge-driven method for mineral prospectivity mapping. Int. J. Appl. Earth Obs. Geoinf. **21**, 556–567 (2013)
8. F. Herrera, E. Herrera-Viedma, A model of consensus in group decision making under linguistic assessments. Fuzzy Sets Syst. **78**(1), 73–87 (1996)
9. C. Kahraman, D. Ruan, I. Doğan, Fuzzy group decision-making for facility location selection. Inf. Sci. **157**, 135–153 (2003)
10. F.E. Boran, S. Genç, M. Kurt, D. Akay, A multi-criteria intuitionistic fuzzy group decision making for supplier selection with TOPSIS method. Expert Syst. Appl. **36**(8), 11363–11368 (2009)
11. Y.M. Wang, T.M. Elhag, A fuzzy group decision making approach for bridge risk assessment. Comput. Ind. Eng. **53**(1), 137–148 (2007)
12. N. Capuano, F. Chiclana, E. Herrera-Viedma, H. Fujita, V. Loia, Fuzzy group decision making for influence-aware recommendations. Comput. Hum. Behav. **101**, 371–379 (2019)
13. Y.C. Wang, T. Chen, Y.L. Yeh, Advanced 3D printing technologies for the aircraft industry: a fuzzy systematic approach for assessing the critical factors. Int. J. Adv. Manuf. Technol. **105**, 4059–4069 (2019)
14. M. Pishdar, F. Ghasemzadeh, J. Antuchevičienė, A mixed interval type-2 fuzzy best-worst MACBETH approach to choose hub airport in developing countries: case of Iranian passenger airports. Transport **34**(6), 639–651 (2019)
15. T. Chen, Y.C. Wang, M.C. Chiu, Assessing the robustness of a factory amid the COVID-19 pandemic: a fuzzy collaborative intelligence approach. Healthcare **8**, 481 (2020)
16. J.M. Merigo, M. Casanovas, The fuzzy generalized OWA operator and its application in strategic decision making. Cybernet. Syst. Int. J. **41**(5), 359–370 (2010)
17. D.F. Li, Multiattribute decision making method based on generalized OWA operators with intuitionistic fuzzy sets. Expert Syst. Appl. **37**(12), 8673–8678 (2010)
18. Z.X. Su, G.P. Xia, M.Y. Chen, L. Wang, Induced generalized intuitionistic fuzzy OWA operator for multi-attribute group decision making. Expert Syst. Appl. **39**(2), 1902–1910 (2012)
19. S. Zeng, Pythagorean fuzzy multiattribute group decision making with probabilistic information and OWA approach. Int. J. Intell. Syst. **32**(11), 1136–1150 (2017)
20. M. Zarghami, F. Szidarovszky, R. Ardakanian, A fuzzy-stochastic OWA model for robust multi-criteria decision making. Fuzzy Optim. Decis. Making **7**(1), 1–15 (2008)
21. J.R. Chang, T.H. Ho, C.H. Cheng, A.P. Chen, Dynamic fuzzy OWA model for group multiple criteria decision making. Soft. Comput. **10**(7), 543–554 (2006)
22. D. Dhouib, Fuzzy Macbeth method to analyze alternatives in automobile tire wastes reverse logistics, in *2013 International Conference on Advanced Logistics and Transport* (2013), pp. 321–326
23. Y. Ozdemir, K.G. Nalbant, Personnel selection for promotion using an integrated consistent fuzzy preference relations-fuzzy analytic hierarchy process methodology: a real case study. Asian J. Interdis. Res. **3**(1), 219–236 (2020)
24. F. Chiclana, E. Herrera-Viedma, F. Herrera, S. Alonso, Some induced ordered weighted averaging operators and their use for solving group decision-making problems based on fuzzy preference relations. Eur. J. Oper. Res. **182**(1), 383–399 (2007)
25. H.M. Lyu, W.J. Sun, S.L. Shen, A.N. Zhou, Risk assessment using a new consulting process in fuzzy AHP. J. Constr. Eng. Manage. **146**(3), 04019112 (2020)
26. Y. Dong, G. Zhang, W.C. Hong, Y. Xu, Consensus models for AHP group decision making under row geometric mean prioritization method. Decis. Support Syst. **49**(3), 281–289 (2010)
27. T. Chen, Y.C. Lin, A fuzzy-neural system incorporating unequally important expert opinions for semiconductor yield forecasting. Int. J. Uncertain. Fuzz. Knowl.-Based Syst. **16**(01), 35–58 (2008)

28. H.C. Wu, Y.C. Wang, T.C.T. Chen, Assessing and comparing COVID-19 intervention strategies using a varying partial consensus fuzzy collaborative intelligence approach. Mathematics **8**(10), 1725 (2020)
29. E. Van Broekhoven, B. De Baets, Fast and accurate center of gravity defuzzification of fuzzy system outputs defined on trapezoidal fuzzy partitions. Fuzzy Sets Syst. **157**(7), 904–918 (2006)
30. T. Chen, Y.-C. Wang, C.-W. Lin, A fuzzy collaborative forecasting approach considering experts' unequal levels of authority. Appl. Soft Comput. **94**, 106455 (2020)
31. H.C. Wu, T.C.T. Chen, C.H. Huang, Y.C. Shih, Comparing built-in power banks for a smart backpack design using an auto-weighting fuzzy-weighted-intersection FAHP approach. Mathematics **8**(10), 1759 (2020)
32. T. Chen, A hybrid fuzzy and neural approach with virtual experts and partial consensus for DRAM price forecasting. Int. J. Innov. Comput. Inf. Control. **8**, 583–597 (2012)
33. M.-C. Chiu, T. Chen, Assessing mobile and smart technology applications to active and healthy ageing using a fuzzy collaborative intelligence approach. Cogn. Comput. **13**(2), 431–446 (2021)
34. G. Zheng, N. Zhu, Z. Tian, Y. Chen, B. Sun, Application of a trapezoidal fuzzy AHP method for work safety evaluation and early warning rating of hot and humid environments. Saf. Sci. **50**(2), 228–239 (2012)
35. F.R.L. Junior, L. Osiro, L.C.R. Carpinetti, A comparison between Fuzzy AHP and Fuzzy TOPSIS methods to supplier selection. Appl. Soft Comput. **21**, 194–209 (2014)
36. T. Chen, Y.C. Wang, H.C. Wu, Analyzing the impact of vaccine availability on alternative supplier selection amid the COVID-19 pandemic: a cFGM-FTOPSIS-FWI approach. Healthcare **9**(1), 71 (2021)
37. J.M. Blin, Fuzzy relations in group decision theory. J. Cybernet. **4**, 17–22 (1974)

Printed in the United States
by Baker & Taylor Publisher Services